# Beyond the Little Blue Book

# Making Our Core Values Our Corps Values!

**BOB VÁSQUEZ**
**CMSgt, USAF (Ret.)**

WAJBOOK PRESS

*Beyond the Little Blue Book: Making Our Core Values Our Corps Values!*



Published in the U.S.A by Wajbook Press, San Antonio, TX.

Paperback Version: ISBN 978-1-7357466-3-0
Kindle Version: ISBN 978-1-7357466-4-7
Library of Congress Control Number: 2021933674

For information about special discounts for bulk purchases or speaking opportunities, please contact us at bobvasquez@bobvasquez.com.

# CONTENTS

# FIRST THOUGHTS

**To know but not to do is not to know.**
**Chinese proverb**

It's 0130 and I can't sleep. I have to share this with you....

October 2005.... I'm teaching my night class at the University of Colorado, Colorado Springs.... We're on break when I happen to look at my Motorola Razr and notice a missed call from my, then, lovely bride of twenty-five wonderful and fulfilling years. (It's 37 wonderful and fulfilling years as I write.) Deb knows not to call unless it's really important. I call her to find out what's up. "Hey, Baby! Are you okay?" I start. "Yeah, but there's a guy named Limoge who says he needs to talk to you as soon as possible. He left his number," Deb replies. I'm trying to impress my students with my techiness so I don't have anything to write with. One of my favorite students, Kathy, walks by so I ask her if she has a pen and paper I might borrow. She does and hands them to me. Deb gives me Limoge's number, tells me she loves me, and hangs up.

I call Limoge. He's a Lieutenant Colonel in the Colorado National Guard. He asks if I'd be willing to talk at his boss's commander's call in two weeks. (He'd invited Chief Master Sergeant of the Air Force, Number Five, Bob Gaylor, who is one of my favorite mentors, but Chief Gaylor is not available but suggests he contact me.) I ask for more specifics. The talk is at noon on Friday of that week. "Sir, I'd love to, but I'm speaking at the NCO Academy Graduation Dinner Thursday night. There's no way I could catch a flight and make it to Denver in time for your event. I'm so sorry," is my first reply. Silence on the other end of the phone. Then he asks, "What if I had a plane there for you to fly you back first thing in the morning, Chief?" Although I don't really know what he's saying I can't say no. "I think that would work, Sir," I tell him. "A C-21 will be there to fly you to Denver. If you leave at 0600, you'll make it to Buckley in time for the event," are his "instructions." "There's one thing though, Chief," he continues. "You'll have to detour a bit and fly down to Dulles to pick up General Fogelman." Retired General Ron Fogelman, the 15th Chief of Staff of the Air Force, is who the colonel is referring to. I've known General Fogelman for many years. He's a great officer, gentleman, and leader. "Of course, Sir!" is my reply.

So it happens just as Lieutenant Colonel Limoge planned. I do my talk Thursday night. I was in flow that night, by the way. Next morning, at Zero-dark-thirty, I'm delivered to the flight line where my plane is waiting for me. And so is Limoge. I'd never had my own plane before so this is very cool. We fly down to Dulles and pick up General Fogelman and head west. During our three-and-a-half hour flight I barrage General Fogelman with questions I've always wanted to ask him. I'd know him for many years, but never had a real sit-down talk with him. Now, he's a captive audience. I'm sure he got tired of my questions. My best question was, "Why did you bring the Core Values to the Air Force?"

General Fogelman brought the Air Force Core Values of Integrity First, Service Before Self, and Excellence in All We Do to the Force by way of "The Little Blue Book" in 1997. They're still our core values as you read this. We had core values before The Little Blue Book was published, six of them, that no one knew or cared about. The Little Blue Book was a succinct way of expressing what our Air Force should be about. I was taught to carry that little blue book with me at all times so that I could refer to it whenever I might have a question about what to do. I was taught to carry it in my right pocket. Right is the position of honor for Warriors in the Profession of Arms.

"Sir, why did you bring the Core Values to the Air Force?" was my question. His reply was very powerful, "Chief," he said, "I was trying to regain our soul." As he said that, I could sense a sincerity and conviction I've sensed only a few times in my life. I could see it in his eyes.

...to regain our soul.... With all due respect, I believe that if we hadn't lost our sole prior to General Fogelman becoming the Chief of Staff, we were on the verge of it. It's not my place to judge so I won't give my opinion on what had been occurring the years before. I hope someone shares that in another book. What I know is that there was something missing in our Air Force. I don't know that anyone, especially we enlisted warriors, could have pinned it down but maybe it was that we were losing our soul.

What you're about to read is about regaining our soul. It's my personal opinion, not sanctioned by anyone else or any institution, but I think, and hope, that the thoughts I share with you will help you contemplate and maybe even help you commit to doing your part to regain our institutional soul. Again, I'm not sure we've lost it, nor am I saying we're losing it, but sometimes, in order to maintain it we need to be reminded, re-blued, if you will, about the foundation we each took an oath to support. As Samuel Johnson once said, "Men more frequently require to be reminded than informed." My entire purpose here is to remind us of what The Little Blue Book expects of us as Airmen in the Profession of Arms and to share a perspective that may be a little different than you've heard before, but that I think will help you commit to doing your part. I'll remind you several times that, although the core values are institutional, it takes individuals to uphold them. Those individuals include you and me.

bob vásquez, zero-dark-thirty, 8 Feb 2015

There's nothing new here. But as Samuel Johnson once said, "People need to be reminded more often than they need to be instructed." My purpose is to help you empower yourself with the values and virtues you need to become a Leader of Character in your own world, by reminding you of what you already know. All I've attempted to do is help you discover, sometimes *RE-discover*, your own stories and thoughts. And to remind you that you can go beyond the ordinary on a daily basis. By doing so you'll affect a whole new generation of leaders.

As I wrote this book, I had the paper copy of the old Little Blue Book in front of me on my desk so that I could refer to it as needed. It's my Old School upbringing, I suppose. As far as I know, that version was never digitized, so the hard copy is it. I also had the new version on my computer, stored in the cloud, to refer to. Things have changed. How we learn and communicate has changed. How we apply the Core Values has changed. It will always be a dynamic process. The people and times have changed. How we perceive what's important to us has changed. But what hasn't changed is the Core Values themselves. They are foundational. General Fogelman was right. They're our collective soul. We can't lose it.

Who, is going to ensure that we uphold our Core Values by practicing them daily, by making them habitual, by ingraining them in our culture so that they become our identity? I am the one! YOU are the one! HEIRPOWER!

# FOREWORD

*"Any time you make a commitment, you sacrifice freedom." - Dr. Arthur Schwartz, Distinguished Scholar, Center for Character and Leadership Development*

When I first heard that statement, I struggled with it. Certainly, I thought, there are commitments that *create* freedom. But the more I reflected on this, the more I bought into it. When we make a commitment, whether it's a promise, contract, marriage or oath, we choose a path. In doing so, we also clearly declare what paths are no longer appropriate or acceptable.

We sacrifice freedom because we accept the blessing of being who we said we would be, and we accept the burden and accountability when we fall short. A commitment is "Yes to this" but also "No to that." When we choose to aim at a target, we choose to not aim at other targets. So, there it is. If you make a commitment, you've got less freedom. That's the deal.

Then why would we make a commitment? Because it matters It's kind of all that matters. Our commitments define us, and signal to the world who we are trying to be as individuals, organizations, professions, and a nation. This is our identity. And nothing more clearly signals our identity to ourselves and others than what we declare as our Core Values.

The Little Blue Book exists to clarify what path we are choosing as members of the Air Force tribe. In three beautifully clear Core Values, we welcome the test to come, and accept the sacrificed freedom associated with living outside those values. We say “Yes” to Integrity, Service and Excellence, and “No” to the easy wrong, selfishness and apathy.

But saying you’re committed is light years away from living it out in the challenging day-to-day, real-world testable moments. To borrow from Mark Twain, that’s the difference between a lightening bug and lightening. It is in those moments that we most clearly feel the freedoms exchanged for this noble identity.

The Air Force has declared the Values, and each member who takes the oath accepts the challenge of proving through their actions if they were serious about what they committed to.

The book you are about to read is designed to clarify and connect you to what is at stake. Using his unique style and humor, Retired Chief Bob Vásquez helps to unpack the importance of the Core Values so they don't remain on the pages of the Little Blue Book. In doing so, he reminds us why the commitment is worth the sacrifice.

So don't just read this book, wrestle with it. Because you had better buy in to the target you're aiming at. Highlight, underline, do the exercises, write in the margins, breathe deeply, and consider how the stories or insights might apply in your life context. Just don't be lazy. The topic – your Core Values, "Our" Core Values – is too important. At the end of the day, your values must be owned, not rented. So do the work.

In the 20+ years Chief Vásquez and I have been wingmen to each other in developing Leaders of Character, I can state without a doubt that "My Chief" (as I call him) is "All In" when it comes to taking Core Values from words to actions. Follow his lead. Make the commitment. It's worth it.

Dr. Kevin Basik, USAF Lt Col (Ret.)

# PROLOGUE

I appreciate the implied power of words. I deliberately selected the word "beyond" for the title of this book because I think we have to go beyond the mere words you'll read, to applying meaning to the thoughts and stories I'll share with you. And as I'll admonish you to consider throughout the book, what we continuously do will create our character, our culture, and our identity. The power we'll create, which I call HEIRPOWER!, will last beyond our experiences and the time we share with those we care about.

The Merriam-Webster dictionary says that "beyond" means "something that lies outside the scope of ordinary experience." How and why Airmen serve is outside the scope of ordinary experience. What every Airman has gone through starting on day one, is EXTRAordinary. As Jimmy Johnson said, "The difference between ordinary and extraordinary is that little extra." Every Airman gives that little extra on a daily basis. The ordinary citizen wouldn't even consider doing the crazy, extra, stuff Airmen do to protect their country and the strangers who send them to war. That same dictionary refers to "ordinary" as "deficient in quality." Let me tell you, EVERY Airman in the United States Air Force is, by their character and willingness to sacrifice for their country, a person of high quality. Every Airman has been vetted extraordinarily to meet and exceed the standards required of a Leader of Character.

What I aim to do with these thoughts and stories is to remind you of who you are and what you represent, the greatest fighting force ever assembled on the face of this earth! Be GREAT, because you ARE!

# THE PYRAMID

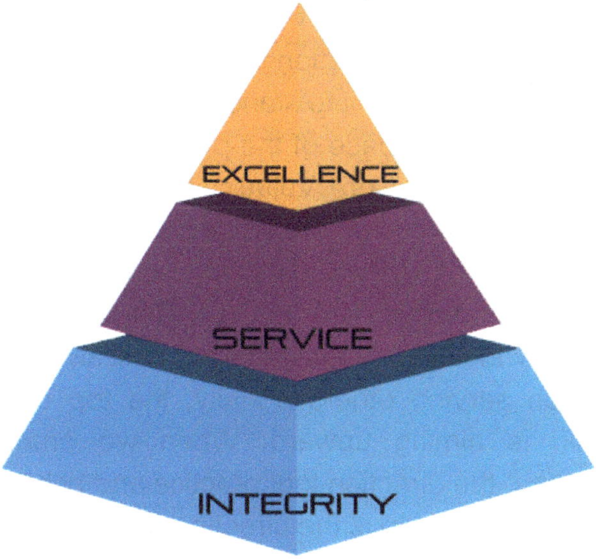


Most of the United States population is made up of visual and kinesthetic learners. There's scientific evidence of that. Images and engagement are the way we learn best. Having said that, the way I intend to engage you in this learning process is through stories that, I hope, will remind you of who you are. To cover the visual aspect, I've included an image of how the Air Force Core Values stack up. In my culture the symbol of the pyramid is sacred. What you learn here will, I hope, be sacred to you in due time. The pyramid is a very powerful symbol in several ways. First, it has a large and solid base. The foundation is strong enough for everything else above it to stand. I think it's apropos that Integrity is the base of the pyramid as it is the foundation for our Core Values. The center, from which much of the symbol's power is developed, is the principle of Service. As we aspire to, and practice, becoming Leaders of Character, the center of who we are and what we do will always be based on serving others. Finally, the top of the pyramid is aiming upward. When we practice Excellence, the direction that practice and behavior will take us is up. Up is the direction we want to go. Ever getting better. The sky IS the limit!

**Do not tell people how to live their lives. Just tell them stories and they will figure out how those stories apply to them.**

**Randy Pausch, The Last Lecture**

**If you talk to a man in a language he understands, that goes to his head. If you talk to him in his language, that goes to his heart.**

**Nelson Mandela**

**Thoughts educate our minds, but stories capture our hearts and souls.**

**bob vásquez**

# THANKS!

There are so many people to thank for inspiring me to write down what I've been living the past fifty years. Thank you, each of you. You've had an impact on me, my family, and everyone who reads this book.

Dr Russ Sojourner (Lieutenant Colonel, USAF, Retired), Dr Kevin Basik (Lieutenant Colonel, USAF, Retired), Dr Jerome Limoge (Brigadier General, USAF, Retired), General Ron Fogelman (USAF, Retired), Dr Renae Skelton (Maj, USAF, Retired), Dr Stephen Covey, Maj Dana Lyon, CMSgt John Evalle (USAF, Ret), Shane Coyne (Lt Col, USAF, Retired), Jessie Brown (CMSgt, Deceased), George Moriarty (CMSgt, Retired), Dr Joe Sanders (Colonel, USAF, Retired), General Steve Lorenz (USAF, Retired), My Dad, My Tíos: Nacho, Raul, and Rumaldo, all of the USAFA Grads that I've been proud to raise in the Core Values since 2002, and my family, Deb, Tesa, and Elyse, who served with me at all of the places we were blessed to live in.

# DEDICATION

"Isn't this a GREAT day to be an American Warrior?" has been my battle cry, mantra, if you will, for decades. It's evolved into a new one, which I'll tell you more about soon. As I've grown older, I've realized the value of living THIS day and I try to make it the best one because it's the ONLY one I have.

This day. It's interesting how we never know what THIS day will bring and the effect it may have. I joined the Center for Character Development (now the Center for Character and Leadership Development) at the Air Force Academy on 16 November 2002. I never, even in my wildest dreams, expected to serve there. NEVER! How I got there was, truly, a miracle.

I met the best officer I've ever worked for, and with (more with), on that day. He was a lieutenant colonel by the name of Russ Sojourner. We'd talked by phone across the Atlantic but had never met. When he offered me the job, and I accepted, I wondered what I'd gotten myself into. But when Deb and I met Lieutenant Colonel Sojourner, we IMMEDIATELY hit it off. It came to pass that he and I would invest many hours, way before normal duty hours, philosophizing and solving the world's problems. I miss those early-morning chats.

We often pondered, but never concluded, how many folks was the minimum number of those we touched to be considered effective. We always defaulted to the easy one, the one. If we connected with just ONE person, that might be enough to affect the Academy, the Air Force, and, maybe, the world.

As you read what I, and some friends, have written here, consider one of my favorite admonitions...I AM the one! Stephen Covey once told me that "the key to the many IS the one." I'm convinced that our society and our culture has deteriorated to the point that most people aren't willing to take personal responsibility for their actions and behavior. It's so much easier to blame someone else for the woes of the world, of OUR world. How does that make sense? It doesn't.

I once read that Gandhi said, "Don't try to change the world, try to change worlds." Here's how I take that thought: We EACH live in our very own world. Even though we live with others, within our family, our workmates, any group of people we interact with daily, we create OUR world by what we think, say, and do. Dr Covey also taught me that "everything is created twice," first in the mind, then in the physical sense or behavior.

Each of us has the capacity to affect OUR world however we choose! The purpose in my writing what you're about to read is that I'm convinced that when we think about the consequences of our choices, we're bound to make better ones. If I do this, that will happen. If I do THIS, THAT will happen.

I'm hoping that reading this book will help you make better choices about what you do and, hence, who you are. It starts with taking personal responsibility. As Warriors in the Profession of Arms, it's incumbent upon each of us to take responsibility and to make the right choices that will lead us to be effective followers and leaders who can, and will, protect our country and our way of life. I AM the one!

Back to my first meeting with Lieutenant Colonel Sojourner. That day, as we talked about what he'd hired me to do, he told me that my job was to "develop leaders of character." "GREAT!" I replied to him, "what's THAT?!" He thought a moment then responded with "I'll get back to you on that." He never did. We cussed and discussed what that might be for most of the year-and-a-half we served together. My perspective was, and still is, that a Leader of Character in the United States Air Force, is, simply, a person who embodies the Air Force's Core Values. Years later, the Center developed a Conceptual Framework for Developing Leaders of Character that, not surprising, is directly aligned with our Core Values. I love it when I'm right. Oh, by the way, the Chief is ALWAYS right! Just sayin....

Character is the aggregate of one's thoughts, emotions, language, and behavior manifested in one's attitude, feeling, communication, and actions in a given moment. Its practice becomes our identity as observed internally and externally.

The key here is EMBODIES! Not just proclaims them. Not just defines them, but EMBODIES those values that are the foundation of our institutional and personal identities. That means we don't just talk about them, we actually apply them on a daily basis. What I hope will come out of this writing is that you will embody those values on a daily basis. YOU are the one!

And back to what Gandhi and Covey said…EACH of us has the capacity to change OUR world for the better. It's our choice. We can't blame someone when we make a bad choice. Oh, I suppose we can, but it's not very effective if we strive to be Leaders of Character. My purpose in writing down my thoughts is to help you think about what's important and to empower you to motivate yourself to make your world, and, eventually, OUR world, a better place.

I'm convinced that my mentor and professor, Dr. Don Hardisty, was right, "Life is simple, once you understand its complexities." We live in a complex world, but what's most important and the way we can affect it in a positive way is simple. It's just three things…Integrity First, Service Before Self, and Excellence in All We Do. The more we practice those concepts and the more effectively we apply them, the better our world will be.

By the way, Sojo (now retired) and I are still brothers and we still mentor each other. Interestingly, my youngest daughter, Elyse, works for him now. He was the best boss I ever had. I watched him practice the Core Values on a daily basis. I'm grateful for our paths crossing and for the brotherhood that we've developed. Love you, Man!

I dedicate this work to you! You are the one who made me the one!

As you read what's ahead, keep in mind that I don't have THE answers. I'm providing you thoughts so that YOU can find the answers. They're within you! Make it so!

HEIRPOWER!
16 Nov 2017

# INTRODUCTION

When I came into the military, I heard people refer to each other as Brothers and Sisters in Arms, but it wasn't until I finished Basic Military Training that I started to feel a sense of family with the men and women I graduated with. We had just endured several weeks of physical, emotional, and mental challenges that helped us bond together, unifying us into a Profession of Arms. Years later, I met retired Chief Master Sergeant Jonathan Harris and he really taught me the importance of strengthening the depths of my relationships with my brothers and sisters who served. Eventually that led me to meeting retired Chief Bob Vásquez who had been a Professional Military Education Commandant who worked with Chief Harris in the 1990s. Chief Vásquez planted seeds that would later lead to my own opportunity to work in Professional Military Education, at the United States Air Force Senior Noncommissioned Officer Academy. I often reached back to him to get perspective on the foundation of reaching the heart of our students and the continual need to impress upon everyone who chooses to take an oath to defend the Constitution of the United States, the reason we must always keep our Core Values at the forefront of our actions, decisions, and behaviors.

That encouraged me to allow my students to witness young men and women choosing to take their own oaths of enlistment, right across the street from our Academy, at the Military Entrance Processing Station in Montgomery, Alabama. It was a great reminder to us all that someone's sister was joining our family. Someone's son was leaving home and starting his journey with the military, to meet new siblings and coaches along the way. In fact, many people who serve in the military have a family member who also served, so it really does become a family business.

One of the reminders I would share with my students is that the young men and women who were leaving their homes to serve our Nation were heading off to Basic Military Training, and several weeks later, many of them would have that proud mom and dad in the stands celebrating their graduation and entry into our noble profession.

In the Air Force, the last thing our trainees do before becoming Airmen is recite the Airman's Creed, and when they do, they aren't whispering, they are so very loud and so very proud. And thousands of people watching share that emotional moment, equally filled with pride! Those young men and women leave Basic Military Training on top of the world, ready to be everything our Nation needs them to be. Those trainees are now Airmen and Guardians, and they are filled with pride and optimism. Then they come out to our installations, and how do we receive them? How do we talk about our heritage and about our core values? Do we instill that same sense of purpose and sense of belonging to the newest members of our family?

We continue to reinforce the Core Values for the same reason they were created. Because we need to be immersed in a culture that practices a set of values that capitalizes on the patriotism and conviction we pledged our oath to, committed to supporting and defending the Constitution of our great Nation.

Our values shape our actions, decisions, and behaviors to do this well. I'm excited for you to be reinvigorated by Chief Vásquez's stories – I'm confident they will help us keep our Core Values in the forefront of our considerations as we continue to lead!

Wendy Thi, CMSgt, USAF
Sr. Enlisted Advisor to the Deputy Assistant SAF

# ALL IN!

I express the idea of excellence with a phrase I use a lot, my new battle cry, if you will, ALL IN!

As Warriors, we need a battle cry! One that will help us win the greatest battle we fight every day…the internal one. I don't know about you, but sometimes getting up in the morning to do, for one more day, what I should do to serve my country, is the biggest battle I fight. My battle cry gets me going and, often, sustains my strength throughout the day.

According to wikipedia, *A battle cry is a yell or chant taken up in battle, usually by members of the same combatant group. Battle cries are not necessarily articulate, although they often aim to invoke patriotic or religious sentiment.*

*Their purpose is a combination of arousing aggression and esprit de corps on one's own side and causing intimidation on the hostile side. Battle cries are a universal form of display behavior (i.e., threat display) aiming at competitive advantage, ideally by overstating one's own aggressive potential to a point where the enemy prefers to avoid confrontation altogether and opts to flee. In order to overstate one's potential for aggression, battle cries need to be as loud as possible, and have historically often been amplified by acoustic devices such as horns, drums conches, carnyxes, bagpipes, bugles, etc.* (I'm a musician, so I can relate to this part very easily. Except the bagpipes. Can't STAND bagpipes. Sorry.)

*"Hooah" is a battle cry used by soldiers in the U.S. Army and Airmen in the U.S. Air Force. Originally spelled "HOUGH", the battle cry was first used by members of the 2nd Cavalry Regiment (United States) during the Second Seminole War in 1841, after Seminole chief Coacoochee toasted officers of the regiment with a loud "Hough!", apparently a corruption of "How d'ye do!" Since WWII, the word has been widely used throughout the US Army and gained a more general meaning of "anything and everything except 'no'".*

*"Oohrah" is a battle cry common in the United States Marine Corps since the mid-20th century. The term means "charge". It is comparable to "hooah" in the US Army and the US Air Force, and "hooyah" in the US Navy. The US Coast Guard uses "Hoorah." It is most commonly used to respond to a verbal greeting or as an expression of motivation.*

*"Hooyah" is a word used in the United States Navy to build morale and signify verbal acknowledgment.*

I've served at the United States Air Force Academy since 2002. I taught my cadets to say Hooah for many years, evolving into HEIRPOWER!, then into ALL IN!

I've been an educator since 1980. Probably sooner. One thing I learned in teaching all of my teaching years is that the best way to educate is through questions. Good questions in particular. Questions, by their purpose, elicit answers. How we answer a question is indicative of our character. Our answers say a lot about who we are. So, my most powerful question is...ARE YOU IN?!

ALL IN! ALL THE TIME!

# CORE VALUES

Words can be powerful. I say CAN because the power of words is inherent in the proper use of them. By proper, I mean, using the right word to express what we're wanting to communicate. Sometimes, we use words as if we know what they mean, but really don't. Since this book is about the Air Force's Core Values, let's start by discussing the words, "core" and "values."

I often do things backwards. I was taught by Dr Stephen Covey to begin with the end in mind, so let me start with the concept of VALUES.

Like I said before, life is simple once you understand its complexities, so let's get to the foundational meaning of that word, "values." If you look up the word, "value," on dictionary.com, you'll find that it simply means "what's important." Simple, right? The complexity exists when we add an "s." Now, you have more than one thing that is important in your life. Chances are, you have more than one value. There are several "things" that are important to you, right?

Notice this...what we call our core values, integrity, service, and excellence, are REALLY virtues. We make them values by making them important. Especially when we embrace and embody them through proactive application.

So, what's important to you? What's your first thought when you wake up in the morning? Why? Because it's the most important thing at that moment! I'm old. My first thought, when I'm not being deliberate in how I awake, is, "I gotta pee!" Trust me, you, too, will have that first thought in due time. Notice that I said when I'm not being deliberate. I've thought about this stuff for decades. My first "deliberate" thought is gratitude. The very first thing I think about when I awake at 0300-or-so every morning is to give thanks to the Creator for blessing me with, one, the health to awaken at all, then for my family and friends. That's what's MOST important to me! Then, I go pee. Now, as important as that first thought is, what really defines my values is what I do. I pray first! Then I go pee. Sometimes the order changes. Again, you'll understand one day. My point here is that we always do what's most important to us at the moment. And we live our life in moments, not hours, days, weeks, etc, but in moments.

Notice, also, that I used quotes above when referring to values as "things." When we're deliberate in assessing and maintaining our values I think we'll find that they're not "things." They're people! Better yet, they're persons. What I give thanks for first thing in the morning are the persons I live and work with. Oh, I'm grateful for a safe place, and, of course, to live in a free country, but they're not my highest values.

Every organization has "proclaimed values." It's what the organization SAYS is important. You've seen them. They're plastered all over the walls of the building you work in, on the business' stationery, even on the website. I've seen a lot of that. But does the organization live up to them? I'm not sure why organizations think that proclaiming their values is enough. The truth is that the people in the organization will either embody... or not... those proclaimed values. The thought seems to be, "Let's tell everyone what we value and we will!" Not necessarily so!

The real values of any organization or team, are the "applied values." What the people in that group truly believe is important. What they actually DO. They will act on those, whether they're on the stationery or not. John C Maxwell says that, "People don't care how much you know until they know how much you care." People, very quickly, will know how much you care because you'll be doing that which is most important. It's a natural law.

Here are a few questions to ponder to assess your values. What's important to you? Who chooses your values? YOU do! No one else! No one can force you to believe that something is more important than another thing, or certain people more than others. I cannot convince you that something is more important. You will choose that on your own. I may influence you to choose something, but the bottom line is that you will choose for yourself.

What happens if you DON'T choose? Someone else will choose for you! You will default to someone else's choice for you. Remember when Mom-n-em used to tell you to hang out with the right kids? Have you commended YOUR kids to do the same? Why? Because if you're going to default to someone else's values it's a lot better if those values and choices are honorable.

Where are your values? They're ALWAYS in your hands! ALWAYS! You will ALWAYS do what's most important to you at the moment. I already said it, but it bears repeating. It's the truth. We live our lives in moments, not hours or days or months or years. It's moments! And at every moment you're choosing what's MOST important and doing it. What do you do when you're really hungry? You eat! Why? Because it's most important at the moment! What do you do when you're really thirsty? You drink! Why? Because it's most important at the moment! What do you do when you have to pee? Why? You catch my drift. Now, you can be deliberate in your choices. You can affect your applied values by choosing the right things to do. The more specific you are, the better.

Identify your values. What do you think is MOST important to you, in your professional life and in your personal life? WRITE THEM DOWN! No, REALLY! Stop reading and write them down. This is key. If you don't write things down, you will lose them. YES, YOU WILL! I used to think that that was a sign of old-age. It's not! It's TMI! There's just too much information bombarding you constantly. Write down what you should remember and affect. What's MOST important to you? Know this. EVERYTHING CANNOT, POSSIBLY, BE MOST IMPORTANT TO YOU! Pare your values down to three or four MOST important things. It may be difficult, but it's possible, and it's, certainly, worth it. Go ahead. Set the book down and write down what's most important to you professionally and personally. You may split them up. It may even be a good idea. I'll be waiting for you when you're done.

Thanks for returning. Now, let's have you CLARIFY your values. What do those things you just wrote down mean? Really mean. Many people write down "family." That's honorable, but what does "family" mean? In the military, and I suppose in most, or all, effective organizations, "family" means our nuclear family, that is, our spouse, children, parents, etc, and it also means our extended professional family, the folks you go to war with. However you define the concept of "family," WRITE IT DOWN! Go ahead. I'll wait.

Again, thanks for doing that and for coming back. Now, PRIORITIZE your values. If you have to choose between the four values you've pared those hundreds you started with down to, which is the VERY MOST important? NOW it gets tough. If you have to choose between yourself and your kids, which is MOST important? Between your family and your work? Which is MOST important? Again, this is not easy to do, but it's worth it. I'll share a story with you soon.

Apply your values. DO IT! Are you good at anything? You probably are. How did you get good? How do we improve on anything we do? Practice!

The more you do something, the better you get at it. Remember the adage, "Practice makes perfect." That's stupid. Practice makes permanent. If you want to apply your values, not the proclaimed ones, necessarily, but what you truly believe is what's MOST important to you, you have to practice them constantly and consistently until they become habitual, until you embody them, until they become your identity, until they become who you are. Remember that you're already applying what's most important to you. But is what you're now doing really most important? Are your current applied values what SHOULD be most important? Are they what you want to be most important? It's ALWAYS your choice.

▲

My favorite sport is basketball. I have the best seats in the house at Clune Arena, where my Fighting Falcons play. I have my own bodyguard there. I'll call him George. We've had those seats since 2004, so George has become a good friend. Hey! He protects Deb and me! He's a good friend.

A few years ago, we missed George for several games. I, eventually, asked his sub about him. She told us that George had had a heart attack but was okay. He'd been recuperating and would soon be back to work.

When George DID return, Deb and I were elated to see his smiling face again. We hugged and greeted each other as only family does. I asked him how he was doing and what had happened. He reported that all was well and that, in fact, he was now on a consistent diet and exercise program.

NOW on a consistent diet and exercise program? "George," I admonished him, "shouldn't you have been doing that all along? Maybe you wouldn't have had a heart attack." He, thoughtfully, agreed. Can't change the past, but he promised he'd continue on the right path now.

What's MOST important? It may sound selfish, but you have to make YOURSELF most important! You cannot give what you do not have. If you don't take care of yourself FIRST, what good are you?

I have two daughters. They're both grown now and have their own families. I'm still, and always will be, their dad, though. If I don't hear from them by 0630 every morning, I go ballistic! Something's happened! Where are they? Are they okay? Why haven't I heard from them? I'm gonna call the cops! I have friends in the FBI!

This particular morning, I hear from Elyse and all is good with her. Then, I hear from Tesa. When I answer the phone, I can't believe what I hear. She sounds like Barry White. Google him. He was a singer. A GREAT one!

"Tesa, is this you?" I ask.

"Yeah, Dad," she answers.

"Are you okay?!" I ask. Isn't that a stupid question? She's obviously, NOT okay. I can hear that she's not.

"I'm sick!" Duuuhh….

At that time, Tesa worked at the Child Development Center at the base. "Call in sick, Mija," I commend her.

"I did call, Dad, but we're not going to be in ratio, so I have to go in."

I don't do math. Ratio has to do with fractions, I believe. I don't do math. I already said that. I don't know what she means. But I do know that she shouldn't be handling little munchkins, YOUR munchkins, when she's ill.

Okay, Ms Manager, what do you do if one of your employees calls in sounding two octaves lower than normal? Admonish them to come in to work? We have a mission, by God! That's the complexities, right? Tesa went to work. Remember I mentioned that you cannot give what you do not have. This is a paradoxical event happening. Within the week, ALL of her kids were sick with her symptoms! She gave them what she had.

Now, Ms Manager, what did you tell Tesa about your values? Her health is not THAT important. But what about all the healthy living programs the organization sponsors and the posters all over the building about how important the people are? Proclaimed vs applied values.

Now, I'm not, necessarily, dissing the people Tesa worked for. I'm asking you to consider what you would do, what you SHOULD do, because those types of situations, the complexities, will arise in your life. Be prepared. Prioritize your values. Maybe let your leaders, co-workers, followers, and family know what they are. In fact, you WILL tell them what they are, by your actions. Your values, remember, are always in your hands.

As I mentioned earlier, John C Maxwell says that "People don't care how much you know until they know how much you care." They know how much you care.

▲

It's 6 July 1992. Nellis Air Force Base, Nevada, right outside of Las Vegas. I'm the Drum Major for the Fifteenth Air Force Band out of March Air Force Base in California. We're part of the training center's change of command ceremony. We're on the flight line. One hundred ten degrees! HOT! BURNING HOT! But it's a dry heat! Dry heat or not, it's BURNING HOT! LITERALLY! This ceremony was the second time in my life that, by the end of the event, my feet were burned from the top and bottom. We wore Corfam shoes. Don't EVER wear Corfams! They're terrible for your feet, I don't care HOW pretty they may look. My feet were burned from the sun shining on those stinking shoes and from the heat generated by the cement tarmac we stood on.

Thousands of Airmen are formed up to see two three-star generals exchange the command flag. If you've ever seen a change of command ceremony you know that the actual change of command takes about two minutes. Assuming that the Chief doesn't drop the guidon, that is. Here's how it goes. The outgoing commander takes the unit's guidon from the Chief and presents it to the presiding officer stating, "Sir/Ma'am, I relinquish command." The presiding officer takes the guidon and hands it back to the Chief. Both officers salute each other then the outgoing commander makes a couple of stepping movements as the incoming commander takes his/her place in front of the presiding officer, and states, "Sir/Ma'am, I accept command." The presiding officer presents him/her the guidon after taking it from the Chief. The new commander gives the guidon back to the Chief and the two officers salute, then face the Airmen. It takes about as much time as it took you to read this. Maybe less.

If you're doing the math, you may be wondering how my feet got burned in the short time that it took to change commands as I just described. Well, here's the rest of the story. We were out in that hot burning sun for about an hour. The outgoing commander had to say goodbye to his Airmen. Luckily, his remarks were short. The presiding officer made some short remarks as well. The incoming commander, however, comma (You ever notice that nothing positive follows "However, comma…."?), the incoming commander must have liked the sound engineer's work because he talked for at least thirty minutes! And guess what he talked about? How honored he was to take command of such a great organization and how great his Airmen were, how he cared so much about them that he was going to do great things for them, etc.… The great thing he could have done was to say, "I love you guys! Let's go have a beer!" Had he done that we would have had a reenactment of the scene in Braveheart where Mel Gibson yells out to his warriors that "They may take our lives but they'll never take our freedom!" AAAAAAGH!!!!!! ALL IN!!!!!!!!!!!!

I share that story to make this point: people KNOW how much you care. As much as that general was TELLING his Airmen how much he cared, he was SHOWING them how little he really DID care, at least about them, at that moment. Remember that I told you our values are always in our hands? Let me say it once more, we'll ALWAYS do what's MOST important to us at the moment.

Think about that and choose what you think is MOST important and practice doing that until it actually becomes most important.

Let me sidestep for a minute and explain a term you'll read plenty through this book. That term is "ALL IN!" I'm convinced that the best way we start a learning process is by asking questions. Questions are powerful because they illicit answers. How we answer a question, I believe, is indicative of our character. I know I've already said that. If we can answer a question with "ALL IN!," that's a declaration of commitment. What's left when we're ALL IN!? Nothing! ALL IN! is about giving 100 percent to what it is we're endeavoring to accomplish. When we're ALL IN! to developing our character, we're bound to get better. If we're going to be better, we have to practice what we aspire to be. All the time. What we practice is what's most important to us at the moment. We can be deliberate in that by knowing what our values are and deliberately practicing them until they become habitual, eventually, our identity. Dr Stephen Covey taught me that "the truly free are the disciplined." We can discipline ourselves to live the Core Values on a daily basis, if we choose to. ALL IN! ALL THE TIME!

Our values are tested constantly. Every day. Often, by the hour. Were you paying attention when I told you that we live in moments? The choices we make in the moment affect us forever. We'll ALWAYS do what's most important to us in that moment.

In the paragraph above, where I suggest you prioritize your values, here's what I mean. I'll use myself as an example. My four highest values are Myself, My Faith, My Family, and My Work. In that order. Now, you may think I'm selfish for putting myself first. Your choice. The truth is that we cannot give what we do not have. You may have read that before. I can't serve anyone unless I'm at full, or as close to full as possible, capacity. The power of prioritizing my values is that I can make better choices when I'm tested. Is life such that I can always make the right choice at the moment? Naw…I'm human. I make bad choices like everyone else. But when the opportunity arises that I have to make a choice, I always make it based on my priorities. I try to choose what I do based on my deliberate priorities. Sometimes My Work will take precedence at the moment. But I've made a conscious choice to make that MOST important at the moment. As soon as I can possibly do so, I go back to my prioritized values list.

We're often concerned with balance. How do we balance our work and personal lives? I think that when we talk about balance, we envision the Scales of Justice. You've seen the picture. It's perfectly balanced, fifty-fifty. Life isn't that way. Balance in our lives is going to require doing more of one thing than the other, in that moment, but regaining our integrity as soon as we can. What's MOST important to us changes constantly. But if we have prioritized our values, we're more likely, than not, to do what we really should, having that list.

The boss walks in and asks you to stay longer than planned. What will you do? What would MOST of us do? STAY! Before I'd prioritized my values, I stayed. It happened many times. I called Deb and told her I'd be late getting home because work was more important. Now, I didn't tell her the last part, but she knew what I was saying even though I didn't. Are you any different? Probably not. Maybe you will be now.

Let's say the boss values you. He probably hasn't asked you to stay later on a whim. There's some good reason for it. He doesn't know that today's your first wedding anniversary and that you had planned to take your bride out for dinner to celebrate, or that your mom, whom you haven't seen in three years is arriving at the airport in a couple of hours and you planned to pick her up. Good bosses make good decisions based on good information. I mentioned earlier that sharing your values may be a good thing. Now, you don't have to get too friendly, but if your boss has at least an idea of what's going on with you, she's more likely to make better decisions, even show you that she values you.

Values. What's important? What's MOST important at the moment?! We all have the freedom to choose. We can be deliberate about how we apply what we choose to be important in our lives.

# CORE

I looked up the word *core* on dictionary.com. That web site said that CORE means “the central, innermost, or most essential part of anything.” As we think about our Core Values keep that definition in mind. Our Core Values are central, innermost, and most essential to our identity and our culture as Warriors in the Profession of Arms. A few synonyms for *core* include *essence*, *heart*, and *center*. It’s probably from the Old French word *coeur*, which means *heart*.

The Merriam-Webster Dictionary site says that to embody something means to personify it, which means to represent as a person or as having human qualities or powers. I’m saying all of this to make a point. We have to EMBODY (I’ve used that word a few times) our Core Values! That is, they have to be in our souls. We have to practice them daily if we’re going to make our force what we all know it should be and, I’m convinced, we all want it to be.

I know I’ve been telling you how to think about what’s most important to you. I’ll stop soon but let me ask you to consider doing one last thing when it comes to values. Those people you listed as most important in your life? How will you show it? I know I'm becoming redundant, but they'll know by how you treat them. WRITE DOWN how you may treat them so that they know you value them. Be mindful, though, that you're looking at them from your perspective. Maybe you're good enough to know them just by observing them, but you'll probably be more effective if you ask them how they would like you to treat them. Just sayin.

Real effectiveness will result when your professional values and personal values are the same. When they're inspired by the right virtues. I mentioned earlier that you might want to list your personal values separate from your professional values. My experience is that, eventually, they'll become one and the same. You may practice a certain virtue differently in different situations and with different people, but it will still be the same general virtue. An example might be love. That's a virtue that I align with all of my values. I practice love in different ways in each of my four Personal Core Values and my Professional Core Values. I show my love for my family a little differently than I do for my followers and colleagues. They probably wouldn't appreciate me hugging and kissing them as I do my daughters and grandkids.

# CORPS

I like to play with words. There's another word I'd like to introduce you to and that's *corps*. Yes, it's a homonym for *core*. But they have different meanings. WAIT! Maybe not! Let's say they have different meanings but the same purpose.

*Corps*, by definition, is "a group of persons associated or acting together." That's what dictionary.com says. Remember what I said about embodying our Core Values? When ALL of us embody what we all agree is important to us as a group, as a corps, we'll produce incredible power. We'll create HEIRPOWER!

The concept of *heirpower* is simple. An heir is a *successor*. If you break that word apart, you're left with *success* and *or*. *Success or*. . . . The alternative to success is failure. Surely, you're not striving to fail. You value success. Most of us do. Power is the capacity to act effectively. Who has that? Everyone! When you help others empower themselves with the right values and virtues, you have incredible power. Not to control them, but to accomplish the mission together. I'll quote my friend, Simon Sinek, who says, "Together is better!"

When we're a corps with the right core, we're unstoppable!

The individual application of our Core Values will lead us, each of us, ALL of us, to develop our identity as a Corps. As we embody, Integrity First, Service Before Self, and Excellence In All We Do, we will maintain our soul. They will help us become who we are.

We eventually become what we continuously do. I think Socrates said that. Practice the Core Values until they become our Corps Values. Do so deliberately and consistently. Make them permanent. It will strengthen our soul.

INTEGRITY
FIRST!

**There is no doubt that absolute integrity is the main principle of leadership. The demonstration of integrity in professional relationships is the singular defining attribute of any (real) leader.**

**General Ronald Fogleman**

**People may be impressed by what you do, but they're inspired by who you are!**

**bob vásquez**

**Humility is not thinking less of yourself, it's thinking of yourself less.**

**CS Lewis**

**The x- factor of great leadership is not personality, it's humility.**

**Jim Collins**

# INTEGRITY FIRST!

First, let's start with the foundational Core Value, the one upon which the others stand, the one Core Value that, when we make it a Corps Value, will make us better Airmen, Warriors, citizens, a more powerful team, an incredible force… Integrity First!

By the way, I use exclamation points a lot! See?! My reasoning for that is that the purpose of the exclamation point is to create excitement. Life is, or can be, so exciting when we do what's right. We can choose to get, and be, excited...or not...! My Chosen Brother, Chappy Watties, once told me that there are two types of people in this world: those who jump out of bed first thing in the morning and exclaim, "GOOD morning, God!" and those who crawl out of bed and moan, "Good God…MORNING…." I choose to be the former. I choose to get excited about today!

I was introduced to the concept of integrity by a man who, although we weren't together long, had a deep impact on me and, subsequently, on a lot of others with whom I've shared his teachings. Chief Master Sergeant (deceased) Jessie Brown was a great musician, leader, and mentor. I don't remember first meeting him, but I remember his lesson on integrity as if it happened yesterday.

I served with Chief Brown in the Philippines, as a member of the Air Force Band of the Pacific, around 1978. We often traveled by C-130 throughout the Pacific Theater. I remember the first time I traveled with the Band. The young Airmen loaded the aircraft before the more senior folks got on the plane. I was a young Airmen then. We loaded the plane, then got off of it through the back to get our personal gear. As we walked toward the front to enter the plane for the trip, I noticed that everyone was hanging around outside, waiting. Not sure who everyone was waiting for, I asked one of the veteran NCOs why everyone was waiting to board. "Chief Brown," was the answer I received. Curious as to why we would all wait for the Chief, I asked a teammate why we would do that. I was told that Chief Brown was afraid of flying. He believed that if the plane were to crash, the tail would be the safest place to be, so he would sit in the very last seat as far back of the plane as possible. Everyone, including the officers, would wait for Chief Brown to board first so that he could take the last seat. I was impressed that everyone respected him so much. On this particular day, he had to take care of some business at the squadron, hence, his tardiness. Once he arrived, and he boarded, everyone followed suit.

For some reason, unbeknownst to me, Chief Brown took me under his wing and mentored me without my asking. I didn't appreciate it as I should have until many years later.

We were on a flight across the Pacific Ocean on an occasion when he chose to teach me the concept of integrity. Remember that this was 1978, decades before the three Core Values were introduced to the Air Force. The air crew used headphones to communicate with the cockpit throughout the flight. Those headphones were connected to the cockpit via a very long cable. The Crew Chiefs would often let us listen in to what the pilots up front were saying. It was a very cool experience. On this occasion, after I'd been on the headphones, Chief Brown came up to me and said to me, "Bob, if you're ever on those things and you hear the pilots say that the plane is about to lose its integrity, grab a parachute…there are only eight!" I kinda got his gist, but not really. He went on to tell me that integrity has to do with wholeness. If the airplane is losing its integrity, it's falling apart. Best to grab a parachute ASAP!

When we're asked to define or describe the term "integrity" we usually state that integrity is doing the right thing when no one is watching. We were taught that in Basic Training. That's an important part of how we should see integrity. But who's to know whether or not we've done the right thing when no one is watching? We are, of course, but it takes a lot of personal strength to do what's right every time. I'll share more thoughts on the idea of perfection when I share my stories on excellence but let me tell you right now: NO ONE IS PERFECT! No one does the right thing EVERY time! Dr Stephen Covey once shared an idea with a group of us that seems to make sense to me. He said that life is a continuum. We'll never be at the peak of that continuum because we're human. As long as we're going in the right direction, however, striving to do the right thing every time, especially when no one is watching, we're better than if we're going backward on that continuum. That seems reasonable to me. We'll fail at maintaining our integrity, but as long as we're trying to live that way, we're better. I fail plenty. I know because I'm tested often.

I live in Monument, Colorado, a few miles north of the Air Force Academy. I live with rich people. All of my neighbors make a lot more money than I do. I'm okay with that, it's just the way it is. There's a point to this, just hold on.

We bought our home in Monument because there weren't many people there in 2002. That's changed considerably in the past decade. So many more people. Unimproved infrastructure. Anyway, a few years ago, the intersection of I-25 and Highway 105 was "upgraded." Someone, in their infinite wisdom, decided to put five, FIVE I tell ya, stop lights in that area. I drive west on Highway 105 every workday and I have to cross all five intersections to get onto I-25.

I'm convinced that those five stop lights have radar built into them that know when I'm coming up to them. You know there's a sequence to stop lights, right? Green… amber… red… green… amber… red…. No matter where those lights are in the sequence, they ALWAYS go to red when I come up to them.

Green... amber... red... green... RED! It happens EVERY time! Now, living in integrity is a choice. I can choose to stop or not. I go to work when no one is on Highway 105. We call it zero-dark-thirty. It's still dark. I'm the only one on the road. Rich people don't even know that zero-dark-thirty exists. They get up around ten-ish, when the sun is up.

What's the purpose of intersection stop lights? Safety, right? To keep people from crashing into each other. There's no one out there, except me! No one! It's not ten-ish, it's zero-dark-thirty. If the purpose of those lights is to keep people from crashing into each other, but there are no cars out, other than mine, why can't I run the red lights? Makes sense, doesn't it? Whenever I pose that question to folks, they quickly rationalize that there may be cars out there that I can't see. I can see far enough each way so that there's no possible way a car could come up on me without me knowing it. I've done this a million times. I know what I'm talking about.

It's a test! I ask again, why can't I run the red lights? INTEGRITY! I have to embody it if I intend to be a Leader of Character. More deeply, I can't expect my cadets to do the right thing when I don't do it. But who would know, you ask? I would know. And what if one of those cadets asked me if I'd done the right thing this morning? I'd have to be honest and admit I deliberately failed. I CHOSE to not do the right thing.

I always stop, by the way. I don't like it and I mumble stuff under my breath, but I stop. It's the right thing to do. I've done it so often that it's now habitual. That's a good habit to have developed, don't you think? Doing the right thing when no one is watching? The key in this story, I think, is being deliberate. Being deliberate takes some thought. Let's pretend we all want to make our team, family, whatever group we work or live with daily, better. Wouldn't it start with thinking about how to do that? And wouldn't it all begin with deliberately doing the right thing, whether or not anyone is present? Doing what's right when no one is watching is, clearly, important. What's just as important, maybe even critical, if you aspire to be a Leader of Character, is doing what's right when EVERYONE is watching.

People may not do what you tell them to do, but they WILL do what YOU do, so be careful and deliberate about what you do. As General George Patton said, "We're always on parade."

It's the very first Black Thanksgiving! You know, when stores started opening up with HUGE (Yeah, RIGHT!) sales Thanksgiving evening. It's now become a tradition for my daughter, Elyse, and me to go fight the crowds that evening, even if we don't have anything to buy. I go just to be with her.

Anyway, Elyse invites me to go with her to Walmart this particular Thanksgiving evening. I love her so much that I accept. Actually, there IS something I want to buy, because it's "half price!" I never did the research to see what the original price was, but the ad said "half price" so I'm in. ALL IN!

Here's what was bumfoozling, the sale starts at 2300 (11 PM for you non-military folks.). Got that. But Walmart is open 24 hours. How's THAT gonna work? Not very well, really, read on....

We figure there will be a lot of folks taking advantage of all the great sales tonight, so we'd better get there early. If nothing else, we can get in line. I LOVE getting in line. I've served in the Air Force since 1972. I KNOW how to get in line. Sometimes, at a store, I'll see a line and get in it. I don't have to know what the line is for. It's a line. I've been trained. I get in it. It's a habit.

SO, we get to Walmart about 2200 (10 PM.) As expected, there are hundreds of people throughout the store. And there are lines. I'm like a puppy on a leash. Let me get in line, PLEASE!

Oh, don't let me leave you hanging. The item I wanted to buy that was "half price" was a trampoline. My daughter, Tesa, has two daughters and I thought it'd be a great gift for them. It actually was. And it was "half price!" Where's the line?!

Elyse heads to find her own line (She's been in the Air Force ALL of her life. Literally.). She's watched me get in line all of her life. So, she, too, knows how to do that. I'm trying to figure out where they might have trampolines. AHA! Let me ask. I notice an already-frazzled manager walking by, so I stop her and ask her. She sighs, one of those "I wish I wasn't here" sighs and tells me to follow her. She takes me way back to near the tire sales in the corner of the store. She points to this big pallet of boxes under a sign that says "Trampolines." As I turn to thank her, she proclaims to me, and everyone around, that the sale starts at 2300. Please don't take the boxes until then.

What do people do when they're told not to do something? What they were told not to do! That's exactly what happens. The manager leaves and people start putting trampoline boxes on their carts. Now, let me tell you, in case you've never bought a trampoline. If you have, you know what I mean. Those boxes are heavy, Man! I'm not sure what the actual weight is, but they have to be several hundred pounds. People are struggling to put their boxes on their respective carts and the carts are barely able to hold them up. And, of course, they don't fit INSIDE the cart. You just kinda set the box on the cart just to roll it over to the cashier. (BTW, all of the cashier stations are manned for this event! If you shop at Walmart, you know what a surprise that is.)

I'm watching all of these folks put trampoline boxes on their carts and I get to thinking that I better do the same, before they're all gone. And it seems to be the American thing to do. I'm ALL IN!

As I huff and puff, attempting to get this incredibly heavy box onto the cart I "borrowed" from someone who wasn't paying attention to their cart, hence not protecting it, and you know, you snooze you lose, I hear a voice like out of the heavens ask, "How you doing, Chief?!"

Time stops. Like in the movies. I look around and everyone has frozen in their steps. I look toward the source of the voice only to see a couple of my cadets. At the time I'm teaching at the United States Air Force Academy's Center for Character and Leadership Development. Every cadet has to go through my class. It's a graduation requirement. Which means that they all know me. "How you doing, Chief?"

"I'm fine," is my reply. "How are you?" "We're good. Whatcha doing?" Okay, so what am I going to tell them? "I'm violating what the manager just admonished us are the rules for tonight's sale. THAT'S what I'm doing!" Okay, Okay, I think it, but, luckily, I know the difference between outer voice and inner voice. I keep it inner voice. "Just trying to make my granddaughters happy," is my lame reply. "Oh," they say as they look at each other, then at me, obviously, not sure what I mean. "Well, have a good evening," they smile, as they keep looking at me.

Look. I'm on a fixed income. This item is "half price." If I don't take it now, I may not get one. You'd do the same thing, wouldn't you? Don't judge me. We're no different. WAIT! This book is about being different! The topic here is integrity. Doing what's right when EVERYONE is watching.

Folks, they're watching! Always! Our followers, our students, our kids. They're watching us. It's how they learn. It's how WE learned. If we're going to be Leaders of Character we have to live up to, and value, integrity in everything we do. We have to be ALL IN! ALL THE TIME! We have to model what we want others to learn from us. What they see is what they'll be.

What really stinks is that I teach the Air Force Core Values to every cadet who attends the Air Force Academy. And, I'm pretty sure, they trust me. How can I admonish them to do the right thing when I don't? I'm not that good an actor.

I choose to do the right thing. I put the trampoline box back on the pallet. And I become the sheriff. It's fun, actually. I stand there and, in my command voice, tell folks who are about to put a trampoline box on their cart that the sale won't start for another half hour or so and that they are not to take them until then. It's amazing what people will do, or not, when you act with authority. Oh, I get plenty of dirty looks, but I'm used to that. I'm a Chief!

Finally, 2300 arrives, I put one of the trampoline boxes on my cart, I look over to see my cadets, but they're gone, and I pay "half price" for my item. The hardest part is getting that stinking box in my Jeep. As I'm about to walk away, that stressed-out manager comes over and thanks me for maintaining control of that pallet, which was empty by 2305, by the way. Kinda makes me feel good to do the right thing.

Integrity is doing what's right when no one is watching AND when EVERYONE is watching. Someone is ALWAYS watching! Do the right thing anyway. We learn by watching. Make sure that what people are watching you do is BEING integrity.

Man, did I ever have a Significant Emotional Experience when I asked my, then, five-year-old granddaughter, Nieves, why she did a particular thing. I'll forever remember that almost-confused face as she replied, "Because YOU do, Opa!" What they see is what they'll be. Doing what's right when no one is watching is important. If you intend to be a Leader of Character, doing what's right when everyone is watching is critical!

By the way, a Significant Emotional Experience is an event that, often serendipitously, empowers you see things differently or more deeply. SEE?!

The young men and women whom I've been charged to develop into Leaders of Character since 2002 are, on average, 18 to 20 years old. Ninety-five percent of them are not old enough to drink. As freshmen, they don't get to go back to their homes until Thanksgiving. That's their first time since the last week of June, when they arrived to start Basic Cadet Training. Having integrity when it comes to alcohol, which, by the way, is the catalyst for many problems, can be, and probably is, very difficult to deal with.

Cadet Johnny finally goes back home. After seeing his family, his priority is to hang out with his buds. His buds are not fellow United States Air Force Academy (USAFA) Cadets. They've been attending other colleges. Other colleges' rules, especially for drinking alcohol, aren't nearly as strict as they are at USAFA. At USAFA, if you're caught drinking under the legal age or supplying alcohol to underaged Cadets, you're likely to be thrown out of the Academy.

So, Johnny finds his buds and they go out to have a good time. It seems that, often, a good time includes drinking alcohol. Now, imagine the scene.

Johnny and his buds are out enjoying themselves. Somewhere, soon, someone introduces alcohol to the scene. None of Johnny's buds, nor Johnny, are old enough to drink. Yet, the first thing they do is encourage the entire group to do so. "Go ahead, Johnny, take a sip or two," they say. Again, just imagine how difficult it will be for Johnny to say no. These are the folks he grew up with. He's trusted them most, if not all, of his life. "What's the matter, Johnny? Don't you want to be one of us? Go ahead, take a drink." If Johnny is to display integrity, he has to say no. At least to the drink, and maybe to being one of them. That's hard for a young person to do. But that's what's expected of a Leader of Character. Johnny has to do the right thing when no one is watching and when everyone is watching.

In the new version of the Little Blue Book, titled "America's Air Force," which, by the way, touches on a lot more topics than the original Little Blue Book, states that there are three virtues that demonstrate each of the Core Values. The three virtues that demonstrate Integrity First are Honesty, Courage, and Accountability.

# HONESTY

Honesty, I'm sure we'll all agree, has to do with being truthful. Telling the truth, especially if it will put us in a bad situation, can be difficult. It's often easier to tell someone what we think they want to hear than what they need to hear...the truth. We'll also all agree that outright lying is the opposite of honesty, but not telling the whole truth is also being dishonest.

My daughter, Tesa, was a tween when she came up to me one day and asked if she could do something-or-other. I told her to ask Mom first. She took off and returned a few minutes later. She asked me, again, if she could do that something-or-other. I asked her if she had asked her mom. She said yes. I gave her permission to do what she asked to do. Soon, afterward, Deb came in and asked me if I'd given Tesa permission to do what she was now doing. I told her I had. Uh-oh! What I had failed to ask was, "What did Mom say?" Uh oh! I was in trouble! I should have thought to ask, "What did Mom say?" but Tesa should have told me Deb said no. Not telling the entire story, the whole truth, may not seem like lying, but it certainly isn't honest. We both learned a few things from that Significant Emotional Experience.

Of equal importance to being honest is the willingness to accept the truth. As leaders, we have to build trust within the people we serve with to ensure that they're not afraid to tell us the truth. Sometimes, as Colonel Jessup said in the movie, *A Few Good Men*, "You can't handle the truth!" A real leader, a Leader of Character, has to handle the truth.

One of my favorite bosses, Colonel Joe Sanders, once told me, "The worst thing that can happen to a commander is when the Chief stops talking to him." What he meant was that when the top enlisted person doesn't feel that he/she can be honest with the boss, the boss, and probably the unit, is doomed to fail. Aldous Huxley once said that, "We shall know the truth and it shall make us mad." Better to be mad and fix the problem than not to know that the problem even exists.

When the team has a strong sense of trust, the people will be willing to find and fix problems without fear of having failed. When something goes wrong and the colonel calls you to hold you accountable, the best that can happen is that you can report to the colonel that you're already working on correcting the problem. That requires a climate where honesty, truthfulness, and trust permeate the culture.

# COURAGE

Being honest takes courage. Doing what's right often takes courage. My gut tells me that most of us know what the right thing to do is. Here's one of my favorite adages: To know, but not to do, is not to know. Like most virtues, the more you practice doing them, the less courage it takes. It becomes habitual.

We usually separate courage into physical courage and moral courage. I'm thinking that the physical courage, the doing of something courageous, starts with moral courage, the thinking of doing something courageous if called to do so. I've been honored to know, or at least meet, some real heroes who have done some incredibly courageous things, and when we've talked about courage, those heroes have each told me that they were either doing what they were trained to do or what needed to be done. They didn't have time to think about the possible consequences of putting themselves at risk.

I hope that you are never put in a life-or-death situation where you have to practice the physical courage that our Wounded Warriors or Purple Heart recipients or Medal of Honor recipients have been in, and I surely don't want to disrespect their sacrifices and fearlessness, but, for most of us, the tests of our courage will be seemingly small. And the more we are tested, the more we'll be willing to be courageous in more substantial circumstances.

One of the "tests" of courage that occurs to most of us on a daily basis is correcting others. I will tell you that during my tenure teaching at the Air Force Academy, that's one of the topics we, Permanent Party, talked plenty about. *Why won't Cadets correct each other?* Whether on uniform wear, mutual respect, sexual assault, you name it. I came to the conclusion that they wouldn't because we wouldn't. You can't teach character and leadership, you SHOW it!

I'm at the Air Force Academy graduation. I'm honored to be invited to sit on the stadium floor with all of the academic professors. I'm not sure why I was ever invited to do so, but I do so with honor. As I walk down to the field, I notice my good friend and fellow warrior, Colonel Gary Packard, standing among about a dozen other full-bird colonels. I haven't lost my eagle eyes yet, so I notice that his right epaulette is sticking up above the collar of his service dress coat. I don't think much of it. His peers will correct it for him, I'm sure. NOT! After looking around and talking with some folks, I, again, happen to look over and see the colonel's epaulette is still sticking up above his collar. I can't believe it! And I can't help myself. I go over, salute, say hello to the colonel and let him know I'm about to fix his epaulette. As I do so, he whispers to me, "This is why I need a Chief around me." Now he looked perfect. I'm honored to call Colonel Packard a friend and colleague. I'll ALWAYs take care of him. Now, that may not have been brave or fearless of me to do, and it really didn't take much courage. I've been correcting Airmen for decades. But for a new Airman, especially a Junior Airman, it would probably take courage to correct a senior officer. My point is that to be courageous in large things, we have to start the growth process in small things. Integrity is about doing what's right regardless of the degree of the risk.

You may remember the cheating scandal that occurred at Malmstrom Air Force Base in 2014.

You can read up on the details of that event, but what I'm convinced of, based on my talking with folks near the incident, is that the Airmen charged thought, and the culture supported the thought, that it was no big deal to cheat. Everyone did it. At least, that was the perception. No one, up to that point, had had the courage to even attempt to stop what everyone knew was the wrong thing to do. Courage in small things empowers you to display it in big things. And the more you do the small things right, the more it will become your identity.

My granddaughter, Nieves, and I just finished having lunch at a fast-food place. She's about five at the time. As we go out the doors, some people are coming in. We raised Nieves to practice common courtesies as character traits. She opens the door for them. You know how loud a five-year-old can be when, maybe, they shouldn't be, right? As they walk right past her, Nieves turns around to me and in her seemingly loudest voice proclaims, "Opa, they don't have character. They didn't say thank you!" Immediately, the people turn back to her and say, "Thank you, Little Girl!" They apologize, but it was too late. They failed the test. I tell ya, our character is constantly being tested, whether we realize it or not. You may have read that somewhere.

Deb and I attended the 2018 Warrior Games Opening Ceremony at the Air Force Academy last night, which was spectacular! The Wounded Warriors who were honored are the epitome of courage. As I said before, those with whom I've discussed the idea of courage account it to doing what they were trained to do or what just needed to be done.

Last night, Vice Chairman of the Joint Chiefs of Staff, General Paul Selva, reminded the audience that a group of people who exhibit courage but are often overlooked are the family members of those Wounded Warriors and those who have given their lives for our country. The families don't sign up for what Warriors do, yet they have to live through, often suffer through, the loss or rehabilitation of their loved ones in a totally voluntary capacity.

I'm convinced that courage comes from caring. The families do what they have to because they care about their loved ones, just as Warriors do. When we care about someone enough, we'll find the courage to do what we need to do to protect them. And whether we do find the courage to act with integrity or not, we'll know. We'll know regardless of whether or not others know. We'll hold ourselves accountable.

# ACCOUNTABILITY

Accountability is about doing what you say you'll do. I don't know about you but being accountable to myself is probably the most difficult challenge I have to fight on a daily basis. I have excuses for everything I decide not to do. I'm so good at excuses that I've numbered them all. "I'm not exercising today because of number 13," I say to myself. I tell you, I'm good at it! I've done it all my life.

I have a "to-do" app on my iPhone where I write down all the things I need to do daily. You know what helps me out a lot? I can change the due date in about three seconds. And I never change those due dates to sooner…always later. I love technology! And I can rationalize why I don't do what I promise myself to do. Dr Stephen Covey taught me an important lesson about rationalizing. He asked, "What do we produce when we rationalize?" His answer was, "Rational…lies." Makes sense, doesn't it?

I wish I had a solution for you, in case you do the same thing. But I don't. Okay, maybe I do. I've found that making a commitment or promise to someone helps me accomplish what I promised. When I make a commitment or promise to someone else, it becomes important to me AND to them, and disappointing others, especially someone I care about, is hard to live with. I'm convinced that we don't always do what's most important to us, but we're more apt to do what's most important to someone who is important to us.

You know who will, without a doubt, keep us accountable to our commitments and promises? Our kids! Especially our grandkids. If you don't have either, you'll have to trust me, but if you have either or both, you'll agree. Here's a parental lesson: NEVER say "Disneyland" and "our family" in the same sentence around your kids because, if you do, you just made a promise to them. And they will hold you accountable for it!

Being accountable is as personal as it gets. I've heard people admonish others to "don't take it personally." Life IS personal! It's your life. No one else's. Whether you succeed or fail is based on YOUR personal choices. Period. Dot. Exclamation point! TAKE IT PERSONALLY! Be accountable for your actions. Don't blame anyone but yourself for what happens to you. Okay, there will always be external influences in your life, but ninety-nine percent of what happens in your life is a choice.

I think that one of the problems we're facing in America today is lack of accountability. No one wants to, or is willing to, stand up and say "I messed up. I made a mistake." We blame the President for everything bad that happens in our lives. Do you know how stupid that is?

My speaking theme for last year was "I am the one!" I traveled the country admonishing my audiences to take responsibility and hold themselves accountable for the choices they make on a daily basis...and learn from them. I'm not sure who first said that we learn most from our mistakes. I'm convinced that's foolish. We learn a LOT from our successes. Now, it takes reflection, maybe more than from our mistakes, but we can learn from our, and others', successes. Isn't that one reason you're reading this book? As my chosen brother, DJ Vanas, says, "Learn from others' mistakes. You'll never be old enough to make them all yourself." And take the time to learn from the good and great things you do and continue doing them. Better yet, do what I'm doing as we speak...share them with others. He who teaches learns twice.

Who is willing to make this world a better place? I AM!

Who is responsible for making this a great day? I AM!

Who is the embodiment of the Core Values? I AM!

Who is the one who will take responsibility and hold themselves accountable for their actions? I AM THE ONE!

The bottom line is that you can't tell someone they need a haircut when YOU need a haircut. You can't tell someone they look fat in uniform when YOU do. Okay, okay, you CAN. But if you want to be a Leader of Character, if you want to be an effective leader, if you want to change the world, you won't do those things I just mentioned. That is, you can't be duplicitous. Duplicity occurs when you admonish someone to do what you don't. As I wrote in my book, *What I Learned From Dad Made Me a Better Man!*, "We cannot teach character. We cannot teach leadership. The best we can do is SHOW it! People will then follow." BE accountable!

I mentioned it before…what's key to being in integrity, to being a real leader, is being deliberate. You can't expect things to just happen. You have to MAKE them happen. And to practice integrity, those things have to be the RIGHT things. And, usually, they're small things. But those small things become big enough to make us who we are. Here's how it works….

I was the Senior Enlisted Advisor to the Commander of the 92nd Air Refueling Wing from July 1994 to July 1996. One afternoon, I was at my desk reading through the mounds of paperwork that just seemed to appear daily. We weren't very tech savvy then, so almost everything was done on paper. I'm nearsighted. I can read close-up without glasses, but I can't see further than about three feet without them. Anyway, I was reading a document when I noticed a figure appear at my door. My office was rectangular in shape with my desk at the far end, about fifteen feet from the door. I couldn't make out who was at my doorway, but I wanted to acknowledge their presence, so I looked up for a second and said, "I'll be right with you." I went back to reading what I was reading as the person at the doorway waited patiently (I thought it was patiently.). As I set down the document, I donned my glasses onto my face. OH, CRAP! It was the Wing Commander, My Boss, a Brigadier General! I'd just told him to wait on me! I immediately did what Warriors do when an officer enters the room. I stood up! I apologized! Luckily, General Voellger was one of the kindest gentlemen I've ever known. He accepted my apology, told me what he'd come to tell me (Not that he was about to promote me!), and we went about our business. Or, as Dr Covey used to say, “busy-ness.”

That day, I vowed to always stand whenever anyone enters my office or stands at the door waiting to enter. I’ve practiced that to the point that it’s now a habit. Let me tell you, it takes me longer, and it’s a bit noisy, whenever I get up these days, but I stand up. It’s the right thing to do. I’m deliberate about it. It leads to integrity. I do it every time. Regardless of rank or status. Everyone deserves my respect and that’s one way I show it.

The Book admonishes enlisted warriors to stand for senior people, in particular, officers, who enter the room. But if you want to be a Leader of Character, you do it because it’s the right thing to do. And people see you doing it. If you do it often enough (always), people will follow your example. I’m not done yet. Here’s the lesson.

I’ve worked with a much more senior officer for more than a decade. He hardly gave me the time of day when we first started working in the same place. He’s been in my office several times. I’ve been in his office, just to say hello and to check up on him, many more times. Every time he comes into my office, I stand for him. Just the other day, I walked into his office and he stood up and shook my hand. For no reason. I was just saying hello. The ONLY way to be a Leader of Character is to model what a Leader of Character does. And all that means is to embody our Core Values. Practice them until they become yours. Until they're habitual. Until they become your identity. Whatever you do until it becomes habitual will become your identity.

I'm working at my desk at Ramstein Air Base, Germany, just before the Air Force threw me out. (I did not retire; that would require a voluntary action.) My, then-teenaged daughter, Elyse, is sitting in a chair next to the door. As the Support Group Superintendent, I work closely with several officers, one of whom turned out to be my favorite lieutenant in the whole world. (I'll just call him LT so that I don't embarrass then-Second Lieutenant Eric Carrano.) As I work, LT comes into my office several times asking for information, direction, money, and so forth. Every time he walks in, I stand up. Elyse watches these interactions for a while before asking me, "Dad, why do you stand up every time the lieutenant comes in here?"

“He’s an officer, Baby, and I’m an enlisted person. That’s what enlisted people do.” (I can tell that she’s a little confused.)

“Aren’t you a big guy in the Air Force, Dad?”

“Yeah,” I answer, wondering where she’s going with this conversation.

“And isn’t he a little guy?”

“Yeah, in a way.”

“So why do you stand up for him? Shouldn’t he stand up for YOU?”

Here’s what I tell her. (Hey, she asked for it, so I give her the sermon.) “An enlisted person is supposed to stand up for an officer when an officer enters the enlisted person’s presence. That’s what The Book says. Well, I’ve been a Chief for about a dozen years, and the book has never stopped me from going beyond it, if you know what I mean. I make it a habit to stand for everyone who comes into my office. I do so even though it’s become more difficult as the knees age. It’s a sign of respect, regardless of rank. The way I see it, Elyse, we all may have different functions, but each of us has the same value. Some would say that enlisted folks stand for officers as a sign of subordination. That's crap! Sorry. No one is subordinate to anyone else. The Creator made us all equal. In fact, one of the charters of a professional warrior is to support and defend the Constitution of the United States, which states that we are created equal. Subordination creates negative energy. Humility creates positive energy. Subordination is maintained by an external source. It's involuntary. Humility comes from inside of us. It's real power because we choose it. I've never subordinated myself to anyone, and I never will.

"I don't stand for the lieutenant," I go on. "I stand WITH him! Did you notice that he stands up when I go to him, too? We don't stand for each other; we stand WITH each other. The combination of my power and his creates a third, much greater, power that will break apart without one of us. I know that my "place" is to stand for him, but my strength is to stand WITH him. I hope that this explanation gives you a sense of my point. A powerful leader is humble. He or she realizes that power lies in helping and supporting others. The more I expand my support, the stronger my leadership power will be because the more people I give to, the more people will give to me. Humility is key to powerful leadership. End of sermon!" She understood.

I hope I've made my point. We all have to work together to accomplish anything worthwhile. There's power in developing Corps Values. Don't make the mistakes others have - learn from them. The more seasoned a person is, the more mistakes he or she has made that you can learn from. Enlisted people are different in some ways than officers. We may have different functions, but we have the same value, especially when we have the same values. And we should all value each other. We are all connected. Strengthen those connections. Make your Core Values our Corps Values.

As Lao Tzu said, "Watch your thoughts, they become your words; watch your words, they become your actions; watch your actions, they become your habits; watch your habits, they become your character; watch your character, it becomes your destiny." Your character becomes your identity. It's who you are, measured by what and how you do.

We moved to Monument, Colorado in 2002, after serving on active duty for almost thirty-one years. I've always been a reader, but I became almost obsessed with reading shortly after we moved to Monument. Books can be expensive. Luckily, Monument has a GREAT library system. It's called the Pikes Peak Library System. I've saved thousands of dollars by borrowing books from the library instead of buying them. (I still have a library of several hundred books, but I, surely, would have run out of room for all those I've read all the years we've lived in Monument.) Anyway, before the library went digital and I had to check out all of the books from the library, physically, I befriended the head librarian, Jean. What a nice lady. One day, she stopped me and asked me, "Bob, are you in the military?" "I served for thirty-one years in the Air Force and I've been teaching at the Air Force Academy since 2002, Jean. Why do you ask?" "Because," she replied, "when you walk in here, you don't walk in, you march in. And when you stand there waiting for me to serve you, you stand at attention." "I'll take that as a compliment," I responded. "Take it as a high compliment, Bob," she said.

We become what we practice. Make sure to practice the right things. Remember what I told you about practice. Practice makes permanent. Practice integrity, and you'll, eventually, become integrity. You will become what you do. Be deliberate about what you do. Don't default to what others might do, especially if it's not core-values-driven. BE integrity!

The Profession of Arms Center of Excellence produced a video in which the narrator says that "Doing the right thing when the world is watching takes courage. Doing the right thing when no one is watching takes character." I love that!

Practice honesty, courage, and accountability until it becomes your character. Your character, remember, becomes your destiny and your identity.

# WHOLENESS

Doing what's right when no one is watching or when everyone is watching, honesty, courage, and accountability all add up to the part of integrity Chief Brown was referring to and we often neglect or overlook when we talk about integrity. The root word of integrity is integer. An integer, if you remember high school math, is a whole number. Doing all of those things I just shared with you make us who we are. NOT doing them creates a black hole in our soul. I don't know physics very well, but I know that a black hole has the power to suck everything into it and out of us. NOT living in integrity is like that. It sucks our spirit and identity into an abyss. You ever tell your mom or dad a lie? How'd that make you feel? If you fessed up to it, you're probably okay now, but didn't you feel a great sense of relief when you did? And if you never did there still may be a hole in your soul. Fix it now!

Wholeness is the element of integrity that, again, we usually neglect, overlook, or take for granted. It may be the most important tenet. Being honest and courageous, and being accountable, are only as good as you are whole. My Toltec mentor, Don Miguel Ruiz taught me that, "You cannot give what you do not have." If you're not whole, you cannot give what's required of you. Remember my story about my daughter, Tesa, having to go into work when she was terribly ill? How effective do you think she was at work that day? Not very. You know that.

Chief Brown's admonishment to grab a parachute if the plane lost its integrity really hit me when I read the Little Blue Book for the first time. It stated that "integrity is the ability to hold together…." That's exactly what Chief Brown was referring to… holding together. His reference was to the plane holding together. When I understood beyond his mere words, and as I grew as a leader, I realized that it also means that when we're honest, courageous, and accountable, we're whole and we hold ourselves together. When we all do those things regularly and habitually, we hold the team together. Those virtues come from our core and eventually affect the corps. As my friend, Lieutenant Colonel Rob Marshall, taught me, "When the going gets tough, the tough stick together." The Core Value of Integrity FIRST requires us to take care of ourselves first. Practice those tenets until they become habitual. Until they become your identity.

# HUMILITY

I was honored to work with the person who developed the new Little Blue Book, Lieutenant Colonel (now retired) Dr Kevin Basik. The moment I got a copy and read it, though, I immediately emailed him to complain that a critical virtue had been left out of Integrity First…HUMILITY. It was in the 1997 version. And it was important, at least back in the day. In reality, it's still important although we seldom discuss it. But we expect everyone to practice it.

Maybe we avoid discussing humility because it's very difficult, in fact, near impossible, to assess or measure. Think on that. How would you measure humility? How would you assess that someone is humble? I think you'd say by how they do things. But have you ever done something in all humility that someone considered selfish or even arrogant of you?

My teammates and I traveled to Denver to hear Dr John C Maxwell speak. We were early so we were just hanging out in the theater where he would speak. Dr Maxwell came out of the back and sat in the front row, probably collecting his thoughts.

All of my teammates got excited and started talking about how they should go up to him and introduce themselves, and maybe chat with him. While they were talking, I went up to him and introduced myself and chatted with him for a few minutes before he excused himself to the green room. After his talk, I asked some of my colleagues how they thought Dr Maxwell had done. Several of them told me they thought he was arrogant. WHAT?! NO WAY! I found him to be very humble, especially while he and I chatted before his presentation.

If we're going to measure humility it can't be on our terms. As Pope Francis said, "Who am I to judge?" I'm convinced that we're the only one who can measure our humility. You know when you've done something in a selfish way or in all humility. It's the first definition of integrity…doing what's right when no one can judge you.

# WHY IS INTEGRITY FIRST?

Why is integrity FIRST? It's the foundation on which trust is built and the other two Core Values stand. Remember, integrity is about doing what's right regardless of who is watching, and it's also about wholeness. We can't expect to serve others if we're not whole ourselves and if we don't, at least, strive, to do what's right. We expect each other to excel and do the right things, not the wrong, or bad, things. I'm this close to suggesting that integrity shouldn't be FIRST, that it should be ALWAYS. Integrity ALWAYS! The problem with that idea is that we're not, nor will any of us ever be, perfect. It just won't happen. But as I mentioned earlier, as Dr Covey suggested, the key is to continue on that positive continuum on our daily adventure we call life, not just our military life, but our life in general.

I may have mentioned it before…the way we lead is through modeling. People will do what we do, or not, when they respect us and emulate us.

One of the things I worked hard at doing while serving as the Senior Enlisted Advisor to the Commander of the 92nd Air Refueling Wing was to be present among our Airmen. I was everywhere I could possibly be, as often as I could possibly be there. My intent in doing that was for our Airmen to understand that my general, their wing commander, whom I represented, cared about them and was interested in what was happening in their lives.

On one occasion, the general had heard that some of the crews were not practicing integrity when they were TDY (on temporary duty away from the base). Many of our Airmen were constantly getting in trouble. I decided to go out on a "milk run" or two to see what was going on. Upon my return, I reported to the general that, in my experience with them, the crews were perfect ladies and gentlemen. Being the wise gentleman that he was, he suggested that they were perfect because I was there with them. Now, let me tell you that I was, and still am, far from perfect, although I do my best to live a life of integrity. His point, and the truth is, those crews acted with integrity because their Chief was there. They acted like ladies and gentlemen because they respected my rank, my position, and, I hope, me.

I've said it before, but let me say it one last time, we're always being tested. Our followers, professionally or personally, are always watching and emulating us. You may have heard that “If you don’t stand for something, you’ll fall for anything.” Stand on integrity. Be deliberate about it. Do what’s right ALWAYS, or, at least, as often as you can possibly do so.

Here’s a story from my Chosen Brother, Dr Kevin Basik.

**INTEGRITY IS A CHOICE!**

In an effort to increase strategic outreach in the Center for Character and Leadership Development, we were excited to support a workshop for corporate executives visiting the Air Force Academy. The team knew this was an opportunity to share what we were doing, and to learn from this visiting top-tier leadership team. It was right in the wheelhouse of the vision for innovation and had done the hard work of getting approval from all typical nay-sayers and created content we knew would be high-impact for all involved.

We were excited to explore topics like trust, accountability, culture, leadership and purpose in a new and fresh way. I was particularly excited to introduce them to the research behind a concept called behavioral integrity -- the alignment of a leader's espoused values and their actions (think "walk the talk"). I'll admit, I was pretty geeked out for this. So often, our discussions with leaders involved the "home team" -- cadets, faculty, staff, and military brass who understood our culture, but also represented the same culture with which we were already familiar. I was fired up for this new conversation.

The day arrived, the room was set, the presentation materials were distributed, and the visitors had arrived. There was great energy, and before the kickoff, I could tell this would be a great event to be a part of.

We welcomed the guests, and I took the floor. Two minutes into the day, the phone call came. A colleague answered the phone, stood up, walked to me at the front of the room, and said, "It's Jen, it sounds serious".

My wife was calling from the car on the way to the hospital. Our youngest daughter (6 years old) had shared a snack with a friend, not knowing it contained peanuts. We didn't know how much she had ingested, or if the anaphylactic response was coming (She's an epi-pen carrying peanut warrior!), but the concern was definitely in my wife's voice. At the time, there was no confirmation she'd had an allergic reaction, but there was also no confirmation she was in the clear.

All the preparation and potential for this event, all the excitement for the discussion faded away quickly. Values and priorities flipped into place and made the next ten seconds easy. I simply handed the slide clicker to my colleague and said, "Ladies and gentlemen, my wife and daughter need me, and that's where I need to be. You're in good hands."

This moment was more unfortunate than challenging. It's an easy call when the values being confronted are clear and non-negotiable. For the record, my daughter was fine and didn't need the epinephrine. What was almost as reassuring is how the visiting executives responded. They, too, had invested a lot of time and interest into this event, so one could assume that they would have bristled at the abrupt departure. In the e-mails I received from them later that week to check on my daughter, everyone shared that they were thrilled to see the clarity and quickness in the decision. Some values are universally held, and they would have been disappointed if the investment in the event had created any hesitation.

Kevin Basik, PhD, Lieutenant Colonel USAF (Retired)
Founder and President, Basik Insight, LLC
*Helping create leaders of character in business and life.*

None of us is perfect but keep striving! Remember to continue to move up Dr Covey's continuum. Integrity is the foundation. Everything rests on it, as in the picture I shared with you at the very beginning of the book. But the CENTER of our Core Values is Service Before Self!

SERVICE
BEFORE
SELF!

**Celebrate the differences. Unity is not sameness, it's complementariness.**

**Stephen Covey**

**A person always doing his or her best becomes a natural leader, just by example.**

**Joe DiMaggio**

**We can do no great things - only small things with great love.**

**Mother Teresa**

**Some people spend a lifetime with others and have no impact, some only a short time together but have a lifetime of influence.**

**CMSgt (Ret) Laten Williams**

# SERVICE BEFORE SELF!

Growing up in the 1960s and '70s, people would ask someone who joined the military, which "service" they were in. The idea back then was that people joined the military to serve. Although many Warriors still "join up" to serve, many more than back in the day, join for the opportunities that the military services offer. There's nothing, necessarily, wrong with that. But as I said at the beginning of the book, things have changed. Maybe we haven't lost our soul as a Force, but the purpose many young people have for doing what they do is different than it used to be. Yeah, I'm advocating going back to the future. Our purpose for serving, if we're going to continue to be great, has to be bigger and deeper than ourselves. The new Little Blue Book refers to Service Before Self as more than a job. In fact, it says that "It is not just a job; it is a commitment that takes energy, dedication, and sacrifice. We do not 'work' in the Air Force; we serve in the Air Force." Each of us joins up for our own reasons. I'm convinced, though, that most of us who stay longer than just one "hitch" stay for the same reason, the people.

Service isn't just about doing, it's about being. Back in MY day, men, and, eventually, women, joined the service. Most of the time we didn't know which service. It was the service! And we were bestowed a new, honorable, identity of being a service member. Maybe we need to reconsider the idea that we joined a service and take a new perspective, that we ARE a service?!

I'm going to say something that may offend you but hear me out. Listen to my argument, then you probably won't be offended. Service before self is overrated. Yeah, I said it. Yeah, I know it's blasphemy. But here's why I say it. You cannot give what you do not have. We often think of, and teach, and perpetuate, the idea that Service Before Self is the most important part of being a Leader of Character. It cannot be so. Why? Because we believe that INTEGRITY is first. That's what we say, right? Integrity FIRST. That means, then, that service before self cannot be the most important of our core values. But it is central! (Integrity, Service, Excellence...Service is the center one. See what I did there? Thought you might like that.) You cannot give what you do not have! You have to have Integrity before you can serve. You have to be whole first so that you can give of yourself.

I'm going to assume that you've flown on a commercial airplane at least once. Recall that one of the first things that happens on a flight is the safety briefing usually provided by the flight attendants, or a video these days. There may be an app for that now. Here's my first observation. This is 2022. Doesn't EVERYONE, by now, know how to buckle and unbuckle a seat belt?! My grandkids, the youngest of whom is three years old, knows how to do that. Why, Mr Fight Attendant, are you telling us that?! Okay, I know it's the law. Warriors know how to follow rules and the law. I guess I can accept that.

Eventually, the flight attendant gets to the part about the little yellow cup dropping down from the overhead compartment. First, they tell you, "in the unlikely event that the airplane loses pressure…." Sorry. I wasn't paying attention. I was trying to give the mom with the screaming kids kicking the back of my seat a hint as to how to shut them up so that I can hear what the flight attendant is commending me to do "in the unlikely event…." I say a silent prayer that there are no unlikely events on this flight.

Back to the little yellow cup…. "If it drops down, put it over your nose and mouth and use the elastic strap attached to keep it in place," the flight attendant admonishes us. Okay, another observation. The flight attendant simulates how to put the strap behind the back of your head. They never mess up their hairdo as they model that action. I guess I've seen that enough to realize to let the strap loose behind the back of my head, otherwise, not only will I not have the use of my hands to save myself AND I'll look really silly as the plane plummets. In the unlikely event, of course.

Okay, here's my point...after the flight attendant trains you on how to don that little yellow cup, she will tell you to put it over YOUR face before helping others. A mom would tell you "NO WAY! I'll take care of my kids traveling with me first." It's a natural thing. That's serving others first. See what I mean about INTEGRITY first? You CANNOT take care of your kids unless you take care of yourself FIRST! It goes against everything a mom instinctively knows to do. She's willing to die instead of her kids! Guys probably don't worry too much about it. Okay, not really. We do. Just trying to keep you awake.

If you noticed the pyramid I introduced you to earlier, the SERVICE block is in the center AND it rests on INTEGRITY. You have to be whole, FIRST, if you're going to help, SERVE, others. You cannot give what you do not have. I know I'm being redundant, but I'm trying to emphasize a point here.

Integrity is about being whole, having all of your resources at your call so that you can take care of others.

My favorite saint, Saint Teresa of Calcutta, said that "our work is our prayer." Taking care of yourself does take work. But that work will result in service. In a way, putting that little yellow cup on your face first is the work that you have to do to serve your kids around you.

So, then, how do we serve? Whether we're serving ourselves or others, we have to know who we, and they, are. There are four basic areas in each of our lives that make us who we are. If we know at least a little bit about those areas within ourselves and those we intend to serve, the better we'll be at it. Again, referring to Dr Covey's continuum, we're probably never going to get it exactly right every time, but the more we move toward serving, the better we'll serve. Those four areas that I'm referring to are values, purpose, vision, and influence.

We, first, have to acknowledge that EVERY person, every individual human, has value. The Creator brought him and her into this world for a specific purpose, often unbeknownst to us and even them, but they're here for that. And every person lives in those four areas. We serve when we can help others empower themselves to fulfill their purpose within each of those areas of their lives.

Everyone has values, something that they feel is important to them. As I just said, sometimes they don't know what's important to them. Serving others starts with helping them figure out what's most important to them and helping them empower themselves to do the things that embody those values.

Why do we do what we do? Whatever we do is based on some purpose. It's deliberate. Even when it's the wrong thing. Leaders of Character know why they do what they do, and they understand the value of what they do. They know who is affected by their actions and behavior and strive to do the best for every person they work and live with.

I often admonish audiences that "If you can see it, you can be it!" Vision is critical to success if we're going to continue to be a powerful force. How do I see myself? How do I see others? How do they see themselves? How do they see me? How do we see the future? These are all important questions because they lead us to being who we are, based on how we see ourselves and others.

Finally, how do we influence others as we're influenced by others? Do you influence others? Of course, you do! When? All the time! In everything you do and say, which comes from what you think. That small thing you did today could affect the life of many. The key to the many, you'll recall, is the one. I am the one!

Service isn't always about big sacrifices. Granted, we're honored to have had many Warriors sacrifice hugely, their lives, to be exact. Warriors like TSgt John Chapman, who received the Medal of Honor, posthumously, in August 2018. That's the ultimate sacrifice Warriors make in the line of serving.

I've had the privilege of meeting several Medal of Honor recipients. Interestingly, every one of them told me they were just doing their job, serving, if you will. The circumstances they found themselves in required that they put their lives on the line, but, otherwise, they did what they were expected to do every day.

Service, for most of us, is embodied in small things that have big effects.

I'd just arrived at Hanscom Air Force Base. My purpose in being there was to speak to a group of Chiefs and Company Grade Officers.

My host picked me up at the airport and checked me in at lodging. I hadn't been in my room for a half hour before the phone rang. I don't recall ever getting a phone call this quickly. I didn't know who it was, but, usually, the calls I've gotten in a hotel room are from the front desk. Maybe they needed more information or something of the like. It was Second Lieutenant Andrew Ramsey. Lieutenant Ramsey had recently graduated from the Air Force Academy, where I was teaching. He was a star hockey player for us.

"I can't be at your presentation tomorrow, Chief, but if you don't mind, I'd like to come by and see you," was what he said to me after we exchanged salutations.

"Surely, LT. When?" I asked.

"In about ten minutes," he replied.

"Okay. I'll be here," was my response.

He DID come by and we chatted for about thirty minutes before he had to go back home. He'd ridden his bike. A long way. And he'd ridden it to where I was just to see me. Now, I know that what Andrew did wasn't anywhere NEAR the service a Medal of Honor recipient has offered, but it's service, nonetheless. Chances are you and I won't be asked to sacrifice as our heroes have, but we'll always have a choice of serving in small ways. Andrew felt it was his duty you come see me. I was touched and honored. Evidently, we did something right for Andrew at the Air Force Academy. I count him as a Leader of Character, not just for what he did, but for who he is.

Service, in the Profession of Arms, can be huge sacrifices and it can be small acts of kindness. Most of us will have many opportunities to practice the small acts of kindness. I pray that you'll never be in a situation where you have to choose to risk life and limb. But to make the right decision if you are facing that dilemma, practicing small acts of service, I'm convinced, will prepare you for that.

# DUTY

Both Little Blue Books refer to Duty as the starting point for practicing Service Before Self, so let's start with that. What does "duty" mean? I like one of the definitions that dictionary.com provides; "an act or expression of respect." It also says that duty is "a moral or legal obligation." By the way, both Little Blue Books also agree that respect is a virtue of Service Before Self. More on that soon.

I'm not going to get into the legal obligation Leaders of Character have to serve. A legal obligation, seems to me, would be a contractual relationship. That does exist, but it's low-level effectiveness, probably more efficiency. I'd rather dwell on high-level effectiveness, which is the moral obligation we have as Warriors. When does that moral obligation, duty, end? We all have duty hours, a workday, a Duty Day, which normally lasts from 0730 (7:30 AM) to 1630 (4:30 PM). Does it end at 1630, at the end of our shift, when we leave the base? Once we're done doing what we're paid to do, does our duty end then? When retreat is played and the flag is lowered, does our duty stop then? When does our duty to the Profession of Arms, to being a Leader of Character end? Come on, we've all heard that we're on duty 24/7, but let's be reasonable. We can't possibly live up to these Core Values all the time! Or can we?

I mentioned the word, effectiveness, above. There's a big difference between effectiveness and efficiency. A manager strives for efficiency. A leader strives for effectiveness. Most people don't know the difference, so let me set you straight.

Efficiency is about getting things done. It usually takes good processes and the right resources, like tools, machinery, time, things that will help produce what you're trying to produce. As my mentor, Dr Stephen Covey, taught me, effectiveness has to do with getting what you're wanting consistently. It, too, requires good processes and resources, but not like things. Effectiveness requires the understanding and caring for the people who are expected to produce the product. Dr Covey used to talk about P/PC balance. P being the Product and PC being the Production Capability. (It's all in his book, *The Seven Habits of Highly Effective People*.) Managers have a good idea, if not a definite idea, of the product they're wanting to produce, it's a thing. Leaders, on the other hand, don't really know what the product is. It's a person with unknown capability. I say unknown capability in the sense that an effective leader will help that person grow beyond what either of them think they can accomplish.

Here's an example of the difference between efficiency and effectiveness. When we buy a new car, one of the first things we look for is fuel efficiency. How far will the car go on a full tank of fuel. (I don't mess with electric cars, so I don't know if "tank" is the correct term, but you know what I mean. A full battery?) The further the car can go on that tank/battery, the more fuel efficient it is. Now, effectiveness is measured in this question: does it start? Better yet, does it start and keep going every time I get in it? That's effectiveness. So the car gets 1,000 miles per gallon/battery, but doesn't start half the time. There's no value in that efficiency then. Managers concern themselves with the product, the thing, efficiency. Leaders concern themselves with the production capability, the people, effectiveness.

Now, as it is in life itself, there's a balance to be struck. A good leader practices good management practices or finds a partner who can manage for her/him. And it makes sense to hone your management skills if you're deficient in that. But to get what you want consistently you'll have to practice leading effectively.

The measure of a good leader is that she, or he, doesn't have to be there to be effective. The people, once empowered, will do it themselves. And getting back to duty, once we've embraced the idea that we have an obligation, 24/7, we'll behave in that manner.

Deb and I have the best seats at Clune Arena, where our USAFA Cadets play basketball. We're right at the edge of the tunnel where the players come on and off the court. Basketball's my game!

At this particular game, we're sitting in our great seats, watching the men warm up when this happens. Every team at USAFA has at least one commissioned officer who travels with the team and is their academics advisor throughout the season. Academics are incredibly tough at USAFA so it's a good thing to have scholars with the cadets to help them keep up with their studies. Anyway, Deb and I are watching the men warm up when I notice the full-bird colonel (O-6) who travels with the team, wearing his Mess Dress, coming toward us. This happens to be the same night of the Wing Annual Awards Banquet. I assume he's wearing his Mess

Dress because he's going to the Awards Banquet. In my mind, I commend him for making the time to come see his team before the Awards Banquet. As he approaches Deb and me, I notice that the button that holds his Mess Dress coat closed is unbuttoned. As he gets within hearing distance of me, I call him over. "Sir, can you come over?" "What's up, Chief?" he asks. We've known each other for many years. Now, we're not buds or anything. I know who he is, and he knows who I am. "Sir, your button is unbuttoned," I tell him. He looks up at me and gives me the dirtiest look! An if-looks-could-kill sort of look. He buttons his button and goes out the tunnel.

I'm bumfoozled by his response! I didn't expect him to make a big deal about me helping him out, but I, surely, didn't expect him to get ticked off at me. A little "Thanks, Chief" would have been appropriate.

As he walks off, a security guard, who the colonel walked past on his way toward where Deb and I are seated, comes over and commends me. "Thanks for doing that, Chief!" It would have been inappropriate to say anything, but I'm thinking, "Why didn't YOU stop and correct him? You're a retired Master Sergeant!"

Deb is livid! “Why did you do that?! That was embarrassing!” I’ve just recently realized that when Deb asks me a question, it’s not really a question, it’s a statement. I hadn’t learned that yet, so I go about answering her question.

I take out my wallet and show her my ID card. I ask her, “What does this say my rank is?” She’s now bamboozled by what I’m doing, but she answers, “Chief Master Sergeant.” I go on, “When does it expire?” “It says Indefinite,” Deb replies. “EXACTLY!” I retort. “I’m a Chief! I will ALWAYS be a Chief! My duty, AS a Chief is to take care of people. Officers, enlisted, civilians, families… that’s what a Chief does. It’s my duty to tell that colonel that he’s out of regs. I didn’t appreciate how he took it, but I have to have the courage to do what’s right. It’s what I teach these cadets. I can’t tell them to do what I’m not willing to do. Besides that, if the Chiefs found out I let that go, they’d excommunicate me from the club!” Man, I’m GOOD!

Deb gives me what we call in my culture, El Ojo, the eye. I sit down and don’t say a word the rest of the game. We won, by the way.

My original question was, "When does our duty end?" It doesn't. EVER! I suppose you can argue that once you've left the service or retired, you're not bound to duty anymore. You can argue that. But if we've done our job as a Corps, you'll always be an Airman. A Marine is a Marine for life. He/she may not be on active duty, but in his/her heart he/she is STILL a Marine. I may be a RETIRED Chief, but here's the truth: retired is a status. Chie Master Sergeant is a rank. My retirement orders, which changed my status from active duty to retired, say I will ALWAYS be a Chief! It's my core identity.

In my discussion of Integrity First, you may remember that I commended you to consider that we not only have to be willing to tell the truth, we have to be willing to accept the truth. That colonel should have understood that I was trying to help him out. I wasn't correcting him to make myself look good. I don't need that. A Leader of Character has a never-ending duty to correct others as much as to accept being corrected.

I'm proud to say that I'm a descendant of men who served our country! My Tío Nacho was an

Army Paratrooper in World War II. He was awarded five Bronze Stars. My dad, Pete, served on a submarine in the Pacific during World War II. I lost my 18-year-old uncle, Rumaldo, in Korea. My Tío Rulie served as an Airman in Germany. There was no doubt I had a duty to serve.

Dad was a patriot! No doubt about that! Dad loved to tell me war stories. It's in the blood, you see. Some I have memorized because he told them to me over and over again. I never minded because I sensed his pride in having served. What he remembered most was the people he served with. I can relate to that because the people are the reason we serve. It's certainly not the money! Think on this: those who serve in our military are willing to GIVE THEIR LIVES in the defense of this country.

I can't recall when it stopped, but for at least the first six or seven years of my life, Dad had me convinced that he and I had served in the Navy together. He'd tell me a story and ask me, "Remember that?" I guess he had me brainwashed because I really DID remember! And he implanted the same sense of pride in me that he felt. He taught me how to tie up the legs of a pair of Navy trousers so that in case I fell overboard (There's no water anywhere near my hometown of Deming, New Mexico.) I wouldn't drown. I still remember how to do that. Dad was proud of his country and the opportunities it gave him to serve and to prosper. We were never rich in the financial sense, but we were rich in love and in pride of our country. We gave dad a military funeral. I saluted a lot. Two young sailors folded his flag and presented it to Mom on behalf of a grateful nation. A Color Guard, made up of Veterans, from Silver City, presented him a 21-gun salute and Taps. It was just how he would have wanted it. On behalf of a grateful nation… we thank you for your service, Dad.

# RESPECT

As I said before, both Little Blue Books agree that respect is another critical virtue of Service Before Self.

This is what the NEW Little Blue Book says. "Respect is at the root of the Profession of Arms and bonds every Airman who voluntarily serves. Respect is the feeling of esteem or deference for a person or other entity, but in the Air Force it takes on a greater meaning and importance. Respect is the lifeblood of our profession. Without it, we simply cannot stand strong in the defense of our nation. Mutual respect strengthens our team and eliminates seams that reveal a weakness in the force."

It goes on to say that "Respect in the Profession of Arms goes beyond professional courtesy. It means accepting others for who they are, embracing a heightened personal sense of humility and fostering an environment of inclusiveness in which every Airman is able and eager to offer their skills, abilities and ideas. It means treating the equipment and resources in our possession with care, understanding and embracing the power of diversity and holding those who mistreat others accountable. More than anything, we must respect the humbling mission placed in our hands by the American people, and the impact our weapons and our actions can have around the globe."

I'm convinced that near the beginning, and permeating the entire process of developing our character, lies the capacity to be respectFUL. Developing character starts with a person developing a healthy self-respect that will result in being respectful to other people. Effective leadership requires a respect of, and for, other people. Those who are seen as having character and as leaders respect themselves and others.

Often, people have trouble starting something because they don't know how or where to start. I think that being respectful starts with, as Deborah Norville says in her book titled *The Power of Respect*, discovering what the person you're wanting to respect calls or sees as respectful. It's not enough that we intend to be respectful. We may intend to be respectful and miss the mark only because we haven't been respectful enough to understand where the other person is coming from.

How do we gain respect? By giving it. I worked with a person, Mary, for several years right after I retired from active duty. We became friends and colleagues. She was still on active duty…a lieutenant colonel. She retired about two years after we'd started working together. While she was on active duty, how I referred to her was simple. I called her "colonel." I'd grown to know her well enough to know that she wasn't going to accept me calling her "colonel" after she retired, so I thought and pondered, "What shall I call her? What shall I call her?" I wanted to maintain my respect for her. She thought more liberally than I, so she told me to call her Mary. NO WAY! I'm from the South. South New Mexico. But even there, I was taught to call people Mister or Missus, or Mizz, but NEVER by their first name. So I called her Mizz Mary. All with the intent of being respectful. We worked together another five or six years, all the while with me calling her Mizz Mary.

One day we were chatting, just shooting the breeze, when the topic of how we address people came up. I was freaked out when she told me that she thought people calling her "Mizz," as in Mizz Mary, was disrespectful! I had had to work hard to come up with a respectful form of address for her. And I'd had to practice to do so whenever I saw her.

Now, she tells me, she thinks it's disrespectful?! I never addressed her again. I didn't know how. We still worked together for a year or two before she retired again, but I was sure not to disrespect her.

I decided, after that obviously silly mistake on my part, to do what would, surely, be respectful. Since then, I've asked. Asking someone how they'd like to be addressed is the best way to ensure you're respectful. Now, I still insist on putting a Mister or Missus in front, and I tell folks who expect me to call them by their first name, that I feel respected when people call me "Chief." I feel uncomfortable calling people who call me Chief by their first name, even when they're a lot younger than I. Which most people I know and work with are.

I've heard and even been involved in the argument about whether retired Warriors should be called by their past rank. Why not? They earned it! They're probably still Warriors inside. I don't know any retired brother or sister who served with me, especially for an extended period of time, who wouldn't still sacrifice their lives for their fellow Warriors. If they respect others that much, we should respect them in return.

When people call me "Chief" it reminds me that I'm a Warrior and that others are looking at me for guidance and leadership. It makes me proud to have served and to have achieved reaching the top one percent of the corps. But it also reminds me that if I'm to guide and lead, I have to live up to the expectation of being a Chief. I try as best as I can to carry myself as I did when I wore the uniform. I live by Patton's statement that "we're always on parade" and work at showing my pride in having served with the best people in the world.

My point is that in order to be respectful, we have to know what that means to the person we're trying to respect. It's not about speaking or acting from our paradigm so much as it's discovering, understanding, and behaving in a way that the other person finds respectful. That's where it starts. We often see leadership in the sense of directing. Effective leadership is about empowering. It's about understanding and lifting others to be THEIR best selves.

As Ms Norville says, discover what the person you're wanting to respect calls or sees as respectful by asking them what that is. If you're thinking that you don't have time for that then what you're saying to that person is that you really don't care about him or her. Oh, but I do care, you say. No, you don't. If you care, you'll make time. And one of the key components of effective leadership is caring for your people. Show them you care by making the time to find out what they see as respectful and behaving in that way when you're around them. That's how and where you start to develop respect.

Being respectful is, often, small acts of service. I believe that at the beginning of being respectful is love. I see love as unconditional giving. I learned that from Dad. He modeled it constantly.

I remember one evening, my brother, Rick, brought a friend over for dinner. I believe his name was Joey. Joey was wearing the new faddish pants, jeans that had tears in them. Dad came home. We had dinner. Throughout dinner, Dad kept eyeing Joey's pants. He never said anything to Joey so as not to embarrass him. After dinner, Dad got up, opened his wallet and took all of the money he had in it and gave it to Mom. "Get him a good pair of jeans," he told her. I'm not sure how much money Dad had given Mom, but I know that it wasn't even close to the amount Joey had paid for those jeans.

And I know it was all he had in his wallet. Respect starts with love. And love is unconditional giving.

▲

I've already mentioned that General Ronald Fogelman, Air Force Chief of Staff at the time, introduced the Air Force Core Values via the Little Blue Book, to the Air Force. I've had the honor of knowing the general for many years and I'm convinced that he practices the core values even to this day. He did while on active duty, for sure.

CMSgt Dave Campanale, the Senior Enlisted Advisor to the Commander of Air Mobility Command (AMC), General Fogelman, invited all of the command's senior enlisted advisors to a conference at AMC headquarters to discuss enlisted issues and policies. One of the highlights of the conference was that we'd get to meet and chat with General Fogelman.

As we prepare to meet General Fogelman, then Colonel Art Lichte enters the room we're gathered in and pre-positions us for General Fogelman's entrance and how he'll meet everyone. I happen to be nearest to the door. Col Lichte has us make a semi-circle. The General will enter the room, meet and greet everyone starting at my end and make the circle back to the entrance of the room. As the General enters the room, he smiles at everyone and commences to do what he was told to do, that is, meet and greet everyone starting with me. "Good morning, Bob," he says to me. BUT WAIT! He, immediately stops and says, "I mean, Chief!"

Although we weren't "buds," we'd known each other for a few years. He and Miss Jane had been my guests while I served as the Commandant of the March Air Force Base NCO Academy several times. He and Deb liked to talk about each other's dogs. Anyway, HE knew me well enough to call me Bob, if he chose to. And no one would have corrected him. But at this moment, knowing that everyone in the room was watching and listening to him, he corrected himself and did what is right to do, he called a Chief Master Sergeant "Chief." I always respected General Fogelman and what he did made me respect him even more. All of the Chiefs took note, by the way. We're always watching. By the way, he still calls me Chief. I still call him General. It's the respectful thing to do.

I'm going to go on record and say that General Ron Fogelman was the best Chief of Staff the Air Force ever had. There was no doubt of his loyalty and respect for the enlisted people. He not only talked the talk, he walked his talk. He was, and still is, incredibly respectful of people, regardless of their rank and status. He practices and embodies integrity.

Respect leads to trust. Trust is critical to being a Leader of Character. Sometimes, it's the smallest of things that create the biggest amount of trust. Be aware of whether or not you're being respectful. Based on THEIR perspective, not yours.

General Tommy Franks retired from the US Army as Commander of US Central Command. He was a keynote speaker at a character conference we both spoke at, at the College of the Ozarks. A group of us were invited to step outside the convention center for a photo commemorating the event. I happened to be at the end of the row until General Franks joined us. He chose to stand next to me. As we were being jostled into position for the photo, I dropped some papers I had in my hand. As I reached for them, General Franks beat me to them. He picked them up, handed them to me, and said, "Here you go, Chief." That may not seem like a huge act of respect, but he didn't have to do that.

I would have picked them up or his aide, who, too, moved toward the papers, would have done so. I was moved by this man, of high prestige, who was willing, without hesitation, to help a stranger. Just based on that seemingly small event, I would have trusted him enough to go to war with him. Trust is huge in becoming a Leader of Character.

I mentioned humility in the previous chapter, on Integrity First. Humility, I think, is very closely aligned with Service Before Self. It's about respecting others beyond yourself. Here are a couple of examples of humility that I've been blessed to observe.

One of my protégés, Chief Master Sergeant (now retired) Calvin Williams, invited me to speak at the Air Force Global Strike Command Chiefs' Induction Dinner several months before it was scheduled to occur. Being that the Chief is one of my favorite protégés, I accepted immediately, and we agreed to the future date. Just weeks prior to the dinner, a new Chief Master Sergeant of the Air Force (CMSAF) was selected. CMSAF Number 18, Kaleth O Wright, became the most senior enlisted person in the Air Force not long before the dinner was held. I went to Barksdale Air Force Base, where the dinner was held, not realizing that CMSAF Wright and his bride, Tonya, would be in attendance. When I found out about it, the OC Factor hit me! Oh, CRAP! I was already on my way. If you're wondering what all the fuss is about, this is it. The top enlisted person in the ENTIRE Air Force is going to be listening to an old retired Chief during this dinner. He had every right to tell Chief Williams to tell me that HE would speak instead. The audience may have preferred that, too, although they, including the Wrights, honored me with a standing ovation at the end of my talk. I was privileged to know Chief Wright when he was a Technical Sergeant at the Kisling NCO Academy when we were in Germany together. He was a good man then, and still is to this day. He showed me more respect by not taking over that event than he had to. The way you gain respect is by giving it!

I was honored to meet one of the most humble people I've ever met, President Ronald Reagan.

I served in several Air Force bands (most, unfortunately, don't exist anymore). While serving with the Air Force Band of the Golden West, stationed at March Air Force Base in California, we had many opportunities to play for President

Reagan at various high-level events. Evidently, he liked us because after he retired to the Los Angeles area, he invited us to come to visit him so that he could thank us for what we'd done for him while he was in office. Now, think on that a little bit—a former President of the United States summoning members of an Air Force Band to his office to thank them! Talk about showing us respect!

We took advantage of the invitation and went to his office one bright, sunny day. (It's always bright and sunny in California.) What an exciting time! The President invested more than an hour with us, joking around and telling stories. What a great storyteller! He inspired me to become one as well, as you're noticing. We all had our pictures taken with him. (I have mine proudly displayed in my office if you'd like to come by and see it.) Apparently, his next appointment was some golf tournament. He was dressed casually since he had to go to the course right after talking with us. I was the second-to-last person to be photographed with him. My commander was last. I'd trained him. As we finished taking the photos, one of the President's aides came in and reminded him that he needed to hurry to get to his next commitment. The president acknowledged the remark but finished the photos. He could have apologized and rushed off with his aide, but he respected us too much for that. He stayed with us until we were all done.

I had brought my copy of *An American Life*, the President's autobiography, to be autographed. We hadn't had time for that, however, so as we were being led out of his office, I asked another aide at a desk near the door if she would please ask the President to autograph my book and I would gladly leave enough money with her to cover the postage to mail it to me. Although President Reagan was being escorted out another door, he stopped, having overheard my conversation. He told the aide with him to get my book, signed it, had his aide return it to me, and then went on his way. Did he have to do that? Of course, not. But he did! He cared enough, he respected ME, a kid who came out of a small New Mexico barrio, to stop and sign my book. He even remembered my name and, most importantly, my rank. He signed it…

*To Chief Bob Vásquez—With Best Wishes,*

*Ronald Reagan, May 17 '93*

You're welcome to come see my book as well. I won't let you touch it, but you can see it.

Many people loved President Reagan. Even more respected him. Surely, history will remember what he did for the country, but the people he touched will always remember him for how he treated them. He was an incredibly respectful man. I would have gone to war with him any time, any place. God bless you, Mr. President!

I'll keep saying this: being a Leader of Character is all about the seemingly small stuff we do. Those small things are what make us great and effective. As Maya Angelou said, "I've learned that people will forget what you said, people will forget what you did, but people will never forget how you made them feel."

Here's a story from my mentor and friend, General Steven Lorenz, USAF, Retired.

16 July 2009—SSgt Sarah Price worked diligently as a radar approach control (RAPCON) controller within the 71st Operations Group at Vance AFB, Oklahoma. If anyone had asked her what she did, her response would have been simple and quick: "I control aircraft."

No one would have argued the importance of her job, especially since it directly enables the student flying training that Vance conducts on a daily basis. Had you asked, Sergeant Price would have told you that her ability to make a difference in our Air Force was directly related to her capability to expertly control Vance's aircraft. Her perspective, however, was about to change.

On Sunday, 8 June, at one minute after midnight, the collective bargaining agreement between an Air Force contractor and its labor force at Vance expired, and student flying training stopped. Nearly 800 people went on strike—that's over 40 percent of the base's labor force! These were all key and essential people who not only ran aircraft maintenance operations, but also enabled most base support operations. Suddenly the base was without people to run the child development center, base supply system, environmental management programs, transportation, mail delivery, communication systems, civil engineering operations, and many other services the base's population depends on every day.

The leadership at Vance had to make some immediate changes to reallocate limited resources based on new priorities and realities. Services that had long been taken for granted were now in jeopardy. Student flying training was no longer the most important operation on the base. One minute after midnight on 8 June, Sergeant Price ceased working as a RAPCON controller and became the base's lodging detail noncommissioned officer in charge. When she returned to work, Sergeant Price began training in a brand new capacity to learn the intricacies of military lodging operations and even spent time learning how to handle hazardous waste! She and her team worked every day of the strike, including weekends. Lodging, a base service function with 100 percent occupancy at the outset of the strike, could not fail.

Being an Airman in the US Air Force is all about service. For many, when we start serving in the Air Force, this is simply something we do. As time passes, military service becomes central to who and what we are. This transition happens at different times for all of us, the sooner the better. When the Air Force is who you are, then you have internalized its core values. You'll spend the extra time to finish the job right. You'll stop and pick up the lone piece of trash along the road while jogging. You'll put the needs of others ahead of your own.

Is Sergeant Price's story unique? It certainly isn't in the 71st Flying Training Wing. The entire base population shifted responsibilities to keep the wing running. When the strike eventually ended more than two weeks later, the base was ready to return to its traditional allocation of responsibilities. Sergeant Price, her lodging detail teammates, and the rest of the wing returned to their normal duties.

Student flying training is once again a top priority for Vance AFB. Tranquility is defined by the sound of aircraft engines above. Vance's reallocation of responsibilities during the strike helped the base gain an important lesson in priorities. Base support services provide the foundation for Vance to conduct its primary mission: student flight training. This is just as applicable at every installation across the entire Air Force and not just at Vance AFB. Our daily base support activities can never be taken for granted. From Sergeant Price's perspective, "even the smallest jobs play their part in sustaining the mission."

Stories like Sergeant Price's happen every day across the Air Force: service-focused Airmen helping each other to get the mission accomplished, even if it means moving away from their comfort zone.

In her wildest imagination, Sergeant Price never would have expected to have the opportunity to make such a positive difference in people's lives and in her organization while working for base lodging, not as a RAPCON controller. As Airmen we are all united by our commitment to integrity first, service before self, and excellence in all we do. Through our core values, Sergeant Price made a difference for Vance AFB.

If you ask Sergeant Price today what she does, her response, still simple and quick, sounds a little different - "I am an Airman." You see, serving in the Air Force is who she is. The sooner all of us follow her lead, embracing service and the other core values, the better our force will be.

From *Lorenz on Leadership* by General Steven Lorenz, USAF, Retired

# LOYALTY

One of the most important virtues of Service Before Self is loyalty. When we talk about loyalty the first question has to be "loyalty to what or whom?"

Leaders of Character take an oath to be loyal to protecting and defending the Constitution of the United States. That's a legal obligation. The moral obligation is to maintain our loyalty to the citizens of the United States. We Warriors in the Profession of Arms are expected to give our lives in defense of our country. No other profession asks that. That's what makes us different. And every veteran has made that commitment.

▲

My daughter, Elyse, graduated from the College of the Ozarks in Branson, Missouri. One summer we delivered her to the campus, we found out that they'd just opened the Dick Clark's American Bandstand Theater in town.

You're probably too young to know, but back in the day, all of us youngsters used to watch American Bandstand religiously. It was on television. In black and white. It came on Saturday afternoons. It was our connection to what was hip in the music world. It was hosted by the ageless Dick Clark. Ask your grandparents about the show, and Dick Clark.

So, the Dick Clark's American Bandstand Theater opens up in Branson the same weekend we're there. The headliner is Paul Revere and the Raiders, one of my most favorite groups ever! I MUST see them! I ask Deb, Elyse, and her future husband, Jeremy, to join me. I cop some good tickets and we attend the show.

During the show, right in the middle of it, Paul Revere has EVERYTHING going on on stage that you can imagine, singers, dancers, smoke, fireworks, horses, you name it. As all of these things are happening at once, he stands up from his keyboard and shouts, "STOP!" And everything does. And then he says, "All of you Vietnam War Vets, please stand up." There were half a dozen of us in the audience. We all stood up. And then Paul Revere says, "Welcome home!" He then returns to the show. Sorry. I get a little choked up when I recall this.

I'm okay now. The reason I get choked up is that after the show all of us Vets gathered together. I was the youngest in the bunch. Some of my brothers could barely walk. Several had crutches. Most were naturally stooped over. Some were in wheelchairs. As we talked, everyone agreed that that was the first time anyone of us had ever been welcomed home. Some had returned from that far-away part of the world decades ago.

▲

Let me be honest here. I entered the Air Force in 1972. I didn't join because I was a patriot. In fact, I was trying to run away. I was about to be drafted into the Army. My number had come up. I was going to college at the time with a 4F deferment. I'm not sure why, but all deferments were rescinded, which put me at the top of the list to be drafted. I had a very close friend, Chuck Wilson, who got me an audition for the 15th Air Force Band at March Air Force Base. I was blessed to pass the audition. I received my draft notice about a week after I'd already joined the Air Force.

Many of us served because we had to. I won't pretend I loved it. I didn't. But I served to the best of my ability. At about the three-year point I realized that I couldn't find better people to be around. I joined because of fear. I stayed, and still serve as I write this, because of the people. You probably know what I mean.

As I write this, a great event has occurred. Here's what the internet reported just now....

"An act of Congress honoring Vietnam veterans with a day of recognition was signed into law by President Trump on Tuesday, per a White House press release. March 29 is now designated as National Vietnam War Veterans Day by the Vietnam War Veterans Recognition Act of 2017. The U.S. flag is to be flown in commemoration of those who served in Vietnam."

"Forty-four years have passed since the last U.S. troops withdrew from Vietnam. Due to the unpopularity of the war, veterans returning home were often met with disdain and did not receive support or gratitude for their service."

"Previously, a proclamation by President Obama declared March 29, 2012, Vietnam

Veterans Day. With the Vietnam War Veterans Recognition Act of 2017 signed into law, Vietnam veterans will be commemorated annually for their service."

Welcome home, Warriors, welcome home!

Warriors have a sworn obligation to serve. But there's a group of Warriors who serve voluntarily. I call them the Quiet Warriors. No one embodies loyalty like our Military Families do. No one!

We don't have a holiday in their honor. They just go on doing what we expect of them and reap no real benefits from their sacrifices. I've often said, and have always believed, that the folks who truly have the toughest job to do in the military are the families. They're the Quiet Warriors. Now, don't get me wrong. After more than thirty years of active duty service and seeing all that I saw in those years, I know our Warriors have a tough job to do. But, you know, we Warriors sign up to do those tough jobs. When we go to battle, we know that there's a possibility of not returning. When we get into battle, we do what we're trained to do, and we know what our responsibilities are. A Warrior accepts the fact that, if required, she will give her life for her fellow Warriors and her country.

Military families don't sign up for any of that. The really young families don't know that part of their contract is to possibly live a large portion of their lives without their spouse or without their parent. Oh, we may tell them, but being told and living it are two different things. I recall, while assigned to Ramstein Air Base in Germany, I'd take every opportunity I could to go out to the DV (Distinguished Visitor) Hangar as a C-130, C-5, C-17, or some other type of aircraft would make it's turn toward that hangar. I knew what was about to happen. I'd get out there as quickly as I could. I'd stand there at attention and wait. I'd wait until the DVs in the aircraft were taken off. They called this the Missing Soldier Ceremony. The DVs were Warriors. They were taken off in flag-draped coffins. Luckily, Germany is usually wet most of the year, so I often credited the moisture in my eyes to the weather.

As I'd stand there at attention, saluting my Brothers and Sisters in Arms who made the ultimate sacrifice for that flag that stands for all that's right in this world, I'd think about that little boy or girl whose daddy or mommy was not coming home as expected. What do you tell a three or four-year-old son or daughter who asks, "Where's my mommy?" Or "When's Daddy coming home?" Oh, you tell them the truth as best you can. You tell them they're heroes and you hug them and assure them that all will be well...in time.

I'd also think about the spouse who already had twice the work to do for what was supposed to be a relatively short time, but now would have those responsibilities for life.

Military families have always had the toughest jobs, and they always will. They're the Quiet Warriors. They deserve to be Time magazine's Persons of the Year EVERY year.

We should never forget the sacrifices past warriors have made for us but at the same time we should always remember the service our military families provided those warriors so that they could do what they did. And their work never stops. They continue to be the backbone of the Warrior Spirit.

I was leading a character workshop for my freshmen at the United States Air Force Academy that my second-born daughter, Elyse, was observing. That workshop always started with the Permanent Party Facilitators introducing themselves so that the freshmen could see who was in their presence. I'd have the Facilitators tell the cadets who they were, where they were from, what they did in their service, how long they'd served, and so on. Elyse was the last to introduce herself. As she stood up, she mentioned that she was my daughter and that she'd served in the Air Force all of her life. WHAT?! I'd never thought of that! She was born on an Air Force Base, at Keesler Air Force Base in Mississippi, traveled most of the United States and a lot of Europe as what we call a "dependent." Freaked me out! She was exactly right! She'd served her entire life! Now, as unique as my baby is, and she is, trust me, she's not different than any other of the hundreds of thousands of children who have served in the Air Force and probably haven't ever thought of it. That's service! Let's not forget their contributions. We wouldn't be us without them!

▲

I, vividly, recall Tesa, who was probably ten at the time, being asked where she was from. She answered, without hesitating, "Base housing!"

We can't serve without our families. They deserve huge props. They deserve more than we can ever give them.

I mentioned earlier that the reason people stay in service is the people themselves. Those who serve, Warriors, are a different breed of people than those who haven't. Sorry. It's the truth. I mean no disrespect to those who didn't, but the truth is that Warriors, including the Quiet Warriors, have a different purpose for their lives. Serving for the sake of serving is central to what Warriors do. We stay together, as a Force, because of each other.

I had a Cadet, while at the Air Force Academy, with whom I had a very special bond. Still do, actually. I'll call him Mike. What a GREAT young man! He was everything we wanted in a developing officer. He, however, decided that the Academy, and the Air Force, wasn't for him. I remember those talks we had, him trying to convince me it would be best for him to go somewhere else, me trying to help him see bigger than he could at the moment. He eventually left. He was immediately accepted at an Ivy League school on the East Coast. My love for him has never diminished, but I always wished he'd stayed with us.

A year or so after he left the Academy, I was on the East Coast doing a speaking gig, so I got in touch with him, and we agreed to have lunch. It was great for both of us, seeing each other again. As we were having our lunch, I noticed his eyes tearing up. I asked him what was the matter. He sadly asked, "Chief, can I come back?" "Back to the Academy?" I replied. "Yes, Sir," he said. "I don't think so. Okay, I'm ninety-nine percent sure that you can't, Mike. Sorry. I thought this was what you wanted to do. You have everything you said you wanted. You have your freedom to do what you want. You're a star on the football team. The academic requirements here are nowhere near those at the Academy. Why would you want to go back?" was my long reply to his short question. "Chief, I miss the people. The Cadets I was a part of were the best in the world. These people aren't bad, but their purpose is very different. I know that every one of the Cadets I lived with would have given their life for me. Not here. Everyone is out for themselves, so that they can get ahead of everyone else. I miss the people, Chief...." he finished as we, eventually, talked about other things.

A straight perspective from a youngster who knew. Warriors are a different type of people. They have a different purpose and a different soul. As I said at the beginning of the book, I'm not saying we've lost our soul, but we sometimes need to be reminded of who we are and why we do what we do. The Air Force Core Values are the foundation of our collective soul. They're what make us us.

A huge virtue that leads to service is sacrifice. In the service to our country, Warriors are expected to willingly make the ultimate sacrifice. One of my favorite Air Force Academy Graduates who went on to be a great officer and is now a great person serving as a Civilian Airman, is Dana Lyon. Her story of service continues in the name of her husband, also a USAFA Graduate and a hero in many ways. Here's Dana's story.

## Service Before Self

*"...Dying to self"*

I often say that "Integrity First, Service Before Self, and Excellence in All we Do," are like "Sunday School" answers that just roll off our tongue, often with no heart and soul, or intent behind them. I feel like it's just something you're taught to say when you first join the Air Force; and deep down they really are your values, but it's not like we actually operate and conduct ourselves as if they are the foundation of who we are and why we live.

Service Before Self was a lofty ideal until it became real. It was a value that we, as service members, hold in high esteem, but on 27 December 2013 it was no longer just a value, it became a way of life. Three individuals in uniform — the Commander, the First Sergeant and a Chaplain — entered the room...

*"Captain Lyon, Ma'am, on behalf of the United States Air Force and the Secretary of Defense, I regret to inform you that your husband, Captain David Lyon, has been killed."*

I stood in front of a mirror in a restroom at Bagram Air Base in Afghanistan just minutes after the notification, staring at a shell of myself, completely wrecked, crushed and broken to pieces. I remember asking myself if this was what it meant to live a life of service, of sacrifice. I was at a crossroads in my heart and in my mind, in my faith, and in regard to my military service - either it was all real and worth living *and* dying for, or it wasn't. I determined in that moment, that this was the life I was called to, even though it cost me everything.

Over the last five and a half years, I have processed the meaning behind Service Before Self. I have come to the conclusion that Service Before Self is more than just the willingness to die for your country. It is foundationally and ultimately the willingness to die to self.

Until you determine in your heart that your life is not your own, you simply do not have the capacity to serve others. Laying your life down for someone is an everyday choice to put others' needs above our own. And it doesn't have to be the "ultimate sacrifice," the goal is not to die, but to live by daily sacrificing your desires, surrendering your agenda, in order to lead and love those you are called to serve. That is the ultimate gift of a fulfilled life!

## Humility and Gratitude

I believe this way of life starts with humility. If you aren't willing to recognize you've been given a life that is really only worth living by giving it away, you're missing out on not only one of the greatest gifts to give, but one of the greatest gives to receive as a result. When I pour my heart out and serve others, I receive the blessing ten-fold.

I believe in leading with love and serving others out of the strength you've gained through trials of your past. Strength grows when you've been tested; blessings are birthed from brokenness. Perhaps it is because of what I've experienced in life, the good, the bad and the ugly, that I see life very differently. I see life as something meant to be lived for others, because others have lived and given their life for me and the opportunities I've gotten to experience. Throughout my life, I've seen it modeled in my family and in people I loved and respected, namely my parents and coaches I've had the opportunity to work with.

I see life's challenges as opportunities to love and grow and then pour into those that have been entrusted to my love, care, direction and leadership. That, in and of itself, is humbling - I've been chosen to and get to love and lead some of the most incredible people through my pain and my brokenness. I royally mess things up daily; I'm constantly humbled by grace.

When you see and experience life through brokenness and choose joy instead of bitterness or the victim mentality, you're in the sweet spot of life and thankfulness overwhelms your heart. When you're thankful for what you have, despite the pain, you can live life generously, open handed - enjoying the moments, relationships and the experiences... which, to me, continues the beautiful cycle and reality of Service Before Self. Be thankful for all you've been given - victories and defeats, joys and sorrows, while humbly living your life for others open-handed, ready for whatever life throws at you.

Agape love: self-sacrificing, unconditional love

It's perfect love. This type of love is unattainable, but one who is postured to strive for it can never fail in the pursuit. For in the pursuit, you find life. "For whoever wants to save their life will lose it, but whoever loses their life will find it; self-sacrifice is the way, to find yourself, your true self." Matt 16:25.

Imagine a life where everyone looks out for the interests of others… we would all win! That was what our marriage was like—a constant competition to out-serve each other, and it was the most beautiful picture of love anyone can imagine!

## Service before Self in the Military

Specifically, as a military member, Service Before Self does not simply mean putting the Air Force mission before your needs; it is much deeper. It's the daily sacrifice of your own agenda to serve others, to meet them where they are at, to love, discipline and give your time, energy, and resources making their life better. I believe that "to much that has been given, much is to be expected." It is the trials that build strength in us that allow us, rather beckon us, to use that strength to serve others who cannot do for themselves.

As leaders, we get the opportunity to serve our troops. When what we do on a daily basis becomes our vocation rather than our calling, we start to fear what might happen to our careers. When we take care of our people and serve them, giving them the tools they need to accomplish the mission - even when that might mean just sitting with them and hearing about what is going on in their life, being an outlet for them in the midst of some hard times, they, in turn, will serve us and get the job done for us because they trust us.

## Patterns of Behavior

While Service Before Self is a Core Value of the US Air Force, it's not a one-time decision to all of a sudden start living a life to serve others. It's recognizing the calling that has been placed on your life. As I mentioned before, I thought it was just an ideal and a value I held in high esteem, but, really, it had been woven into my being throughout my life. It might have been a moment in time when I connected the dots after Dave was killed of what the ultimate Service Before Self looked like, but it was not some lofty decision I made for the first time.

The sum total of every decision we make determines our character. Throughout our life we are creating patterns of behavior. We don't wake up and say, "I'm going to be honorable today," or "I'm going to be a leader today." No, we posture our hearts to serve and be used for the betterment of the mission and our people. We wake up giving our best, making the best of every moment, investing in lives of others - day in and day out. It's who we are, it's not a one-time thing.

My athletic journey, while it pales in comparison to the highs of marriage and grief of losing Dave, metaphorically prepared me for the perseverance and resiliency necessary to respond with a heart even more on fire to serve others in the midst of my pain. Athletics taught me so much about responding to setbacks and disappointments. The victories in sport were a result of all the countless hours of training, creating patterns of behavior in my technique. I didn't wake up one day a national champion, I showed up doing my average best, always striving for more.

When I had setbacks, I found a way to overcome. The same could be said about my response to losing Dave, but even more of an eternal impact was seeded in my heart to live for others, using strength through perseverance to love others who have been trusted to my care and leadership.

To achieve beyond your abilities, you must believe in a power beyond yourself. When you realize that God wants to bless you and use you, that He rewards those who earnestly seek him, all psychological limits vanish, and you are free to achieve.

Dana Lyon, Major, USAF

We all have our reasons for "joining up." But joining an organization isn't always the same as serving in it or for it. Sometimes, we're confronted with a Significant Emotional Experience that brings the best out of us. Here's a testimony to that from one of my dearest colleagues, Lt Col (Retired) Shane Coyne.

It was 23 April 2004 and I had just gotten to the Ritz in downtown Colorado Springs where I worked a couple of nights a week as a door guy. The news had just broken that a former NFL football player turned Army Ranger, Pat Tillman, had been killed in action in Afghanistan and I noticed a guy walk by with a homemade white t-shirt that had "RIP Pat Tillman" written on it with a sharpie pen. I remember pausing in reflection as I had heard about this man and how he had walked away from the Arizona Cardinals after the 9/11 attacks, but I was moved to learn more.

At the time, I was a Major in the United States Air Force Reserve and instructor at the Air Force Academy. In the days that followed, I grabbed every news story and article with motivated curiosity and admiration to try to discover what drove Pat Tillman to take his path. As I began unpacking the details, it dawned on me that his was a story that needed to be told to our cadets, our future officers. In fact, I had been in the initial stages of creating a block on one of the Air Force Core Values, Service Before Self, and this appeared to be a perfect fit.

So, I dug deeper and committed myself to learning Pat Tillman's story. I contacted a member of the sports media in Phoenix and was able to get a series of videos that captured much of Pat's journey from NFL player to Army Ranger. Included in this video was an interview that was conducted by the Arizona Cardinal public relations folks. The sportscaster at the CBS affiliate told me it was the only interview Pat ever granted in regard to his intriguing story. I asked him why there was only one and he replied in Pat's own words, "that he didn't want to take the attention and focus off the hundreds of thousands of other members of the US military who were serving every day." What's more, Pat Tillman snuck out the back entrance to the Cardinal's training facility after the interview to avoid other media who had gathered. I stopped and took that all in. Let me get this straight, a person seen in celebrity status who was making millions of dollars is so moved by the 9/11 attacks that he walks away from a life many only dream about, to enlist as a foot soldier in the U.S. Army and doesn't want any publicity. Who does that? I had to know more.

During the interview, they asked him why he did just that. Pat replied that he had always had a great deal of respect for the Flag and what it stood for, and that he felt like he had enjoyed all the freedoms and privileges it stood for. He asked himself "What have I done for my country?" and his answer was that "I haven't done a damn thing". Interestingly, in the duration of that short interview, Pat mentions "the Flag" 4 times. So, I began putting all the pieces together to present to Cadets what I thought would be a powerful piece on Service Before Self. But then, as I continued, I started to see a problem.

Pat Tillman's story is amazing! It's moving! It's inspirational... and it's something that no cadet will likely ever be able to relate to or repeat. I mean, how many of us will rise to the top of our profession and make tons of money only to be moved to throw it all away to step onto the battlefield and put our life on the line for the red, white and blue. The reality sank in. Was this a great idea that will fail to land and connect? I dug deeper. Thankfully, Pat Tillman had much that any cadet, any person, cannot only be inspired by, but moved to action.

In reality, the more I researched, the more I discovered that Pat Tillman didn't see himself any differently than the millions of men and women who had served, bled, and died before him. He had a moral clarity about him that I wished I had. There was no apparent struggle or moral dilemma, nor was there any regret about going from a life of relative ease to put his life, literally, on the line for his country. He saw distinctly that he had a responsibility to step up alongside every other American and he wasn't going to be content unless he did…and that is the beauty of his story. It isn't a comparison about who gave up the most worldly possessions. It's a picture of every single Soldier, Airman, Sailor, and Marine standing up when it counted and their willingness to sacrifice everything they have and everything they are in defense of the place we call America.

But what about this story. It's wonderful and inspiring to cadets but what was its impact on me? The more I shared Pat's story the more it convicted me. The more I got up in front of others, the more I realized that I had something hollow in me that needed to be filled. Pat's quote from the video kept echoing in my mind every single day as I asked myself "what had I done for my country?" ...and I could never get the right answer. You see I had been in the Air Force 13 years and my answer was I hadn't done enough and, unlike Pat Tillman, could easily have. I had no excuse or rationale to the contrary.

It was December of 2004 and the semester was over, and I had some time to sit down and think. The news from out of Iraq was a daily dose of reality that my brothers and sisters were fighting every day and it was then that I knew that I needed to step up. The next day, I told my boss what I must do, and I contacted the Air Expeditionary Force Center at Langley Air Force Base and volunteered to go to Iraq and requested they send me somewhere that would be tough and a place I could make a difference. They called me back the next day and told me I was going to be assigned to a joint task force overseeing all the detainees in Iraq and cleaning up the mess at Abu Ghraib prison in the aftermath of the recent scandal that rocked the value system of the U.S. military.

On Christmas Eve, I shared the news with my parents and sister and after we shed some tears, we prayed, we rallied together, and soon after, I began the long preparations to meet my deployment date departure of 7 April 2005. It was an incredible few months that tested my physical, mental and spiritual strength further than I ever thought possible. I was one of the fortunate ones and returned home safely in the fall of that year.

I will never forget that experience and I will forever be grateful for Pat Tillman, whose decision to step up served as an inspiration to me and countless others who were influenced by his willingness to sacrifice everything for his country. We are all Pat Tillman, and his story shall continue to inspire many today to don our nation's uniform and answer its call for generations to come.

Shayne Coyne, Lt Col, USAF, Retired
USAFA Center for Character and Leadership Development

▲

The Profession of Arms Center of Excellence produced a very cool video titled, Loyalty, in which the narrator states, "you ARE your country. We ARE the Air Force," which reminds me of a thought Master Sergeant (Retired) Lee Richards once shared with me… "We always think of the real Air Force as out there. The real Air Force is in here," as he pointed to his heart. Integrity First starts with knowing, it's using our heads. Service Before Self centers on our hearts, with how we feel, especially about serving others. Excellence In All We Do is about how we act and behave. It's in our hands.

**EXCELLENCE IN ALL WE DO!**

**The strength of the team is each individual member. The strength of each member is the team.**

**Phil Jackson**

**Do you know what my favorite part of the game is? The opportunity to play.**

**Mike Singletary**

**Leadership is a team sport. Even the greatest talent can't win alone.**

**CMSgt (Ret) Todd Simmons**

**The only meaningful measure of a leader is whether the team succeeds or fails.**

**Leif Babin**

# EXCELLENCE IN ALL WE DO!

Excellence is not perfection! You will NEVER be perfect! Oh, you may do something near perfectly, but I bet that when that happens, you'll quickly acknowledge that you could have done it better. Hence, it was not perfect!

As I learned about integrity from my mentor, Chief Jessie Brown, I learned about excellence from another fellow Chief, George Moriarty. He was a Master Sergeant at the time that I met him. He was an instructor at the Noncommissioned Officer Academy at Lackland Air Force Base where I was a student. His presentation of a lesson was so powerful that, although he wasn't my direct instructor, I had to seek him out and ask for his mentoring. In a fifteen-minute conversation, which included the idea of excellence, he inspired me to change my whole philosophy on who I was and how I did what I did. Until that point, I'd been a perfectionist. I was pretty good at a couple of things and I strove to be as perfect as I could at those two things. What Chief Moriarty taught me was that I was bigger than that. I could excel at many things, if I chose to. He taught me that I would never be perfect, but I could be excellent.

Excellence in All We Do requires an ALL IN! attitude. No matter what we do, we should do it to the best of our ability. Our fellow Airmen, all of the people we serve, deserve nothing less. My dad taught me to do my best at everything I attempt. It may not always be enough, but it's all I've got.

Excellence in All We Do begins with knowing what to do. How do we know what to do? We have to practice listening, internally and externally.

Internally, we have to listen to our values, our purpose, our vision, which will all lead to how we're influenced as well as how we influence others.

As I said in the chapter on values, the basic, foundational, definition of the term "values" is "what's important." The internal question is "What's important to ME?" I also mentioned that our values are ALWAYS in our hands. We'll ALWAYS do what's most important to us at the moment. If internal excellence is important to you, then be intentional about it. Be ALL IN! at whatever you do. That doesn't mean perfection, that means give it all you have. It's the least you can do.

Why do you do what you do? What's your purpose? You've heard enough people tell you that if it's all about you, then you'll fail. Not really. We're kidding ourselves when we think that we, anyone, does things totally for other than getting something out of it. Even service is selfish to a degree. It is. Think deeply about it. Even when you're giving unconditionally, there's a tinge of selfishness if nothing more than the good feeling you get from giving. Why should you be ALL IN!? Why give all you have without expecting anything in return? It's our identity! It's who we are (or will, I hope, become). We just do it! It all starts in our core.

You have a choice. You can see yourself as great or you can see yourself as less than you can be. Here's where you'll feel uneasy. I've already admonished you to consider that service before self is overrated. You cannot give what you do not have. I already said that, too. It's okay to be proud of yourself. You've accomplished a lot more than many millions have. You got through Basic Training, for God's sake! Maybe a few times. And if you've been in service a while, you've survived several PSCs! Warriors are the only folks who do both repeatedly. See yourself as excellent, someone who does many things with expertise and professionalism. The word professionalism is defined explicitly in the dictionary. But the spirit of that term is something internal that only experience can fathom. See yourself as a professional and you'll become one, or at least more of one. It will lead you toward excellence.

Are you influenced by external forces? You probably said yes. Most people answer that way. But the truth is that we can CHOOSE whether or not to be influenced. Influencing forces exist around us all the time, but do we go with them, or can we choose otherwise? See what I mean? One of the most influential forces in our culture is the media. Do you believe everything you read and see on television or other social media? That's external. It CAN affect you, if you let it. But integrity requires you to do what's right. To make the right choices and decisions. You can, and probably should, take into account all of the external data you can but, in the end, it'll all come down to the choice you make of how you're influenced.

On the other hand, do you influence others? When? ALL THE TIME! The moment you connect with someone, whether it's electronically or in person, you're immediately influencing them. Malcom Gladwell, author of *Blink*, says that we make up our mind about a person within two seconds of meeting them. That's fast! That's influence! The important question is HOW do you influence others? There are only two ways...positively or negatively. There's no in-between. I'm going to assume that you'll choose the former. How do you influence your followers in a positive way? Each of us, they, have values, purpose, vision, and influence. How are your followers doing in those areas and how can you help them improve? Affecting each of your followers affects the Corps. The first way to find out how your followers are doing is to listen to them. But not just their words. Listen for meaning. Experts say that up to 97 percent of the communication process is non-verbal. Listen with your eyes and your heart as well as with your ears.

I'd just sewed on Chief and had the honor of going back to where I started, March Air Force Base near Riverside, California. I hadn't been there long when my commander, Major Denny Layendecker, summons me to his office.

"Chief," he says, "please, take my office door down."

"I'm sorry, Sir, (I didn't think I'd heard him correctly), did you say take your door down?"

"Yes, Chief. Take my door down," he replies.

"Surely, Sir, but may I ask why?" I ask, bumfoozled.

"I want to have an open-door policy. I want the troops to know they can come in and talk with me whenever they want."

"Okay, Sir," I acquiesce.

Now, as I said, I'd just sewed on my Chief stripes. Chiefs don't do that type of work. They have Duty Sergeants whom they're training to be Chiefs one day. So I find the Duty Sergeant and ask him to take the Major's door down. Yeah, I had to explain why. It was done.

About a month later I'm chatting with the boss and the topic of communication comes up. I ask him, "So, Sir, how's it going with your door down? Are people coming by to chat? You made that announcement in your Commander's Call."

"Chief," he looks at me in a bit of a saddened tone. "Not one person has come by."

“Hmmmm…,” I murmur. “Your intent was to hear their thoughts, their problems, and learn about them, right, Sir?” He agrees. “Here’s a suggestion. How about YOU go out there and hang out with THEM. Where they live and work. (The shop was on the first floor of a dorm. The Airmen lived on the floors above.)”

“I hadn’t thought of that, Chief,” he perks up. “I’ll try that!”

A few days later he comes to my office and excitedly asks me, “Chief, did you know that Carmen is pregnant, and that Jerry’s mom is coming to visit him next week, and that Smitty inherited a lot of money?” he goes on and on.

“Yes, Sir, I know all that.” I reply.

“How did you know all that?” he asks, almost surprised.

“Because I do what you’ve been doing, Sir. I hang out with the troops trying to figure out what makes them tick so when I make recommendations to you, they’re based on what’s important to accomplishing the mission and to taking care of our people. We can't be excellent until we know what that means and how we can do it."

"Thanks, Chief! By the way, can you put my door back up?" he asks.

"Can do easy, Sir," I answer. And then I went and found the Duty Sergeant.

My point, again, is that you can't practice excellence in all you do until you know what excellence is. Then, as a Leader of Character, you have to find the resources, ninety percent of which are people, to practice it. And don't forget that just saying you're practicing excellence doesn't mean you're actually doing it. You have to have SOME type of measure and you have to seek feedback.

I served (indirectly) under General Robert Rutherford while he was the Commander of Air Mobility Command. He used to go around his bases, checking on his people and his bases' operations. He'd get to Base X and as he was being shown around would ask people, "How are things going?" Imagine that. A four-star commander asking how things are going. How do you answer that? "Great, General!" Regardless of the truth, you'd answer in that way. His follow-up question would hit people between the eyes, though. "How do you know?" Duh... crickets... bamboozled looks.... What a powerful question when you're striving to be excellent at all you do, isn't it? How do you know? As I said at the beginning of this chapter, in order to be excellent at what you do, you need to know what to do...and how to measure what you do. If you intend to be a Leader of Character, if you're striving to accomplish something, figure out how you'll know whether or not you're succeeding.

Another important question to consider when you're evaluating your personal excellence is WHO knows? Often, it's not us. It's those we lead. Here's a story that former President of the United States, Jimmy Carter, tells that I think epitomizes the importance of measuring our behavior and considering who knows how we're really doing.

A group of young missionaries complete their training and are dispatched to their first assignment. Upon arrival, as they're driving down a dirt road, they spot an old farmer walking in their direction. Just before they pass each other, the missionaries stop as the farmer stops next to them.

One of the missionaries gets out of the car, says hello to the farmer and asks him, “Sir, are you a Christian?” The farmer takes a small piece of paper and a pencil out of his coat pocket. He writes on the paper, then hands it to the young missionary. “These are the people I live with,” he says, “go ask them.”

In leading, ask your followers. It’s really simple. Effective leading is encapsulated in the thought, “mission first, people always.” Since this book is about leading people and persons, let me admonish you to consider that how we lead, how we accomplish the mission, is different than it used to be. Things have changed. I know that. You do, too. How you must lead has changed. I certainly don’t advocate living in the past, but I think it behooves us to look at the past to learn how to live in the present and how to affect the future.

Back in the day, we were often “trained” by fear. I remember my first assignment with the 15th Air Force Band at March Air Force Base in

California. Within days of arriving, three of us Young Airmen, at that time, Mike Mormon, Mike Paulson, and I, were taken outside by Staff Sergeant Rick Rose, the Duty Sergeant, for him to tell us what our additional duties were. "Airman Mormon, your job is to…." I don't recall, but it was much easier than mine, I assure you. "Airman Paulson, YOUR job is to…." Whatever. Then, "Airman Vásquez, your job is to keep this sidewalk clean." There was a huge eucalyptus tree right next to the Band building. The building and the tree were separated by a sidewalk, which was now my responsibility to keep clean. That tree shed its bark like it was going out of style. That sidewalk was constantly strewn with eucalyptus bark. I noticed it the first time I walked up to the building. Sergeant Rose gave me specific instructions. "Here, Vásquez, take this broom and sweep that sidewalk. I don't want to see any bark on it. Ever." "WHAT?" I thought. Evidently, I wasn't just thinking it, I was saying it, and more. "WHAT? You want me to sweep that sidewalk every day and keep it clean every day, all day long? Are you crazy (I thought I was just thinking.)?" In those days, it was a big thing to have gone to college prior to enlisting. You graduated from Basic Training as a two-striper, an Airman First Class, vice an Airman Basic no-striper. I'd gone to New Mexico State University for three semesters, which made me eligible for the extra stripes. “Look, Sergeant Rose,” I told him, evidently. “I’m an Airman First Class. I went to college. I’m a musician. I’m a bassoonist. I don’t do that!” Again, I was thinking I was thinking, not saying. Man did that set Sergeant Rose off!! I can still feel it…he stuck his finger in my chest and tells me, “Do it or I’ll kill you!” Let me just say that that sidewalk was the cleanest sidewalk on base for as long as I was there.

The 1997 version of the Little Blue Book referred to different types of excellence; personal, service, community, resources, operations, that, combined, could be labeled as mission excellence. Without second guessing the folks who chose to include Mission as a virtue of Excellence in the new Little Blue Book, the truth is that Mission is not a virtue. Sorry. That’s the truth. But the idea expressed is viable and important. The key to the different types of mission excellence, which, interestingly, is the second word of the description, is FOCUS. You can argue about whether focus is a virtue. Go ahead. But focus is closer to a virtue than mission. To achieve excellence in anything and everything, requires focus. It does. You know that. And if you focus on being YOUR best in all of those areas, not, necessarily, THE best, you'll achieve excellence. It's a natural law.

While serving in the Air Force Academy's Center for Character and Leadership Development, I was honored to meet and serve with a lot of great people. One of my favorites was Captain Renea Skelton. Here are her thoughts on excellence.

When Chief asked me to contribute a story on 'excellence' for his book, I struggled with recalling an elaborate memory from my 22-year Air Force active duty career. This is not to proclaim that there were not any, but as a military member for so many years, excellence was something that I just "did" as one of my core values. Excellence was the foundation for many of my actions and decisions because it was something I believed in and lived as a core value.

To alleviate my struggle, I decided to ask my peers, former leaders, and friends how I demonstrated excellence in hopes of discovering an easy answer. Everyone provided examples such as the many deployments, winning prestigious awards, and the various sacrifices sprinkled throughout my service in uniform. I, however, felt something was missing from their responses; something deeper, something bigger than myself. I knew that I did not want the ego-filled stories of my amazing crusades or the life-altering experiences that I encountered; the stories that they were sharing with me. I wanted something more, something different. As I finished a phone call with a close friend asking about how they saw me as "excellent," my 15-year-old daughter, Kelcie, walked into the room and asked me, "Mom, why don't you just ask your children?"

Time immediately stood still as I sat there questioning how I could have missed them, my own family? I felt embarrassed because these were the individuals who have seen and experienced my life with me every step of the way. They were not immune to the military lifestyle. They had the answer.

I smiled and softly asked, "What does excellence look like for you?" I ensured that I did not mention "excellence in me" or "excellence in you." I was intrigued to hear her own direction and definition of excellence. As she stared directly in my eyes, she simply stated, "Mom, you are excellence." It was such a powerful statement; one that allowed us to engage in a deeper discussion that will resonate within me forever.

Kelcie reminded me of the many sacrifices I endured such as missing holidays, birthdays, and school events to defend her freedom so she can continue to enjoy those things that I missed. She reminded me that, as a dual-military couple, we faced challenges that are unique and foreign to most, such as her mother and father being away at the same time and still raising a family from afar. She continued to remark on how, as her mother, I engrained the Air Force core values within our family dynamic because they were not mere theories but values that we embodied. As I sat and listened to her maturity, I noticed that she did not reference once her own sacrifices. I could not help but think about the many times she had to change schools during the middle of the academic year, the close friends she left behind, or the challenges of assimilating into a new school curriculum. She only talked about my excellence and the impact it had on her.

As the conversation ended, Kelcie left a final statement that struck an emotional nerve: excellence was not something I did, but something I am. In her eyes, I was a hero; in my eyes…I was just being me. She saw me in a different perspective than I saw myself. The military was my way of life, it was normal for me. Although normal for her, the definition of excellence resonated in a unique way for her. Somehow, I forgot along the journey, that the military was her way of life as well. Instead of seeking answers from my peers, leaders, and close friends, my answers were within my own family. The family that accompanied me during the 22-year ride. Although they see me as "excellence" they are the ones who truly epitomize the core value. THEY are excellence!

Dr Renea Skelton, Capt, USAF, Retired
www.reneaskelton.com

# MISSION FOCUS

As I said before, Mission, is not a virtue. But the essence of what the new Little Blue Book refers to in the area of Mission is focusing on the right things. It refers to the need to "harness the ingenuity, expertise, and elbow grease of all Airmen." I see that as focusing on how we can rise to becoming and being excellent. I call that empowerment. And the way we do it is by developing and practicing our Empower Hours.

▲

"I've asked hundreds, probably thousands, of people, "What's MOST important to you? NO ONE has ever said "ME!" That answer, of course, would seem selfish, wouldn't it? SEEM, but not so. In my rediscovery of my Toltec heritage, I've learned that "We cannot give what we do not have." In other words, we have to take care of ourselves FIRST! That's the truth. And as leaders, leading effectively starts with us.

As I was reading *The ONE Thing*, by Gary Keller and Jay Papasan, they introduced me to the concept of Power Hours. That's not a new concept, evidently, but I'd never considered it. In their discussion on focusing on our purpose, they mentioned that how we start our day is critical. It sets the tone for the rest of the day. I've known that for a long time, but I like to take things to the next level, which, in this case is really EMPOWER Hours. The value of starting our day off powerfully is in empowering ourselves to be our best.

Power is the capacity to act effectively. Empowering is about enabling ourselves to produce power. We often think and even express that empowerment is an external thing. WRONG! It's totally internal. We empower ourselves, or not. Others may help us, but it's an internal strength. The hard part about empowering ourselves, sometimes, is knowing how to do so. Here are four suggestions.

The very first thing you have to do to empower yourself is to MAKE TIME to do so. You have to get off your buts.... But I don't have time. But it's hard. But I don't know how. MAKE TIME for yourself. If you have to get up earlier than everyone else, do so. No excuse. Although we use the term "hours" it doesn't have to be an hour or more. Twenty minutes of Me Time, invested wisely and purposefully will empower you beyond your expectations.

Whatever it is you intend to invest your Empower Hours doing, WRITE IT DOWN! A thought doesn't exist until you've written it down. Okay, that's Old School. Capture it on your smart phone or your computer. Evidently, there's an app for that. If you don't, you'll lose it. Something will distract you and you'll either forget or neglect it and you'll be no better than before. You want to invest your first twenty minutes meditating? GREAT! Write it down! Better yet, put it on your calendar. You're more likely to do it then.

Thirdly, DO IT! To know, but not to do, is not to know. I KNOW I need to work on me first. I KNOW the value of clearing my mind and starting fresh every morning, thinking good thoughts. I KNOW that that's what meditating will do for me. But I don't! No excuse, Ma'am! Do it! Like any skill you choose, the more you practice it, the better you'll get at it. And if it's an empowering behavior, and you're intending to empower yourself to be your best, just do it!

Lastly, one thing we don't do well, if at all, is measure our progress. If you do, you're special. One of my Empower Hours is invested in reading daily for twenty minutes. No more, no less. I set the timer on my Apple Watch to keep me on track. I've read 24 books so far in the past four months. Now, they're not War and Peace-size, but I've learned a lot, like the importance of Power Hours, which are now Empower Hours. I keep a list of what I've read. Knowing that I'm doing well helps me to continue doing what's working for me. When you consider how you're doing in whatever it is you're trying to improve, if you don't measure it, how will you know?

What I'm telling you is really simple, and you know it already. The hard part is doing it. But you can! I suggest you commit yourself to empowering yourself in the areas we now call Resiliency Pillars, physical, mental, emotional, and spiritual. You can hone each every day for twenty minutes. And you can do it throughout the day if it suits you better. Make time for you! Invest in you! You're the MOST important person in your life. Then, go help others empower themselves. That's truly effective. That's HEIRPOWER! And that's how you accomplish the mission!

# DISCIPLINE

Discipline starts with values and commitment. We're all committed to something, or someone, we really care about. We're likely to actually do what we're committed to doing. If my commitment is, truly, to the Core Values, then I'll probably develop habits that will help me live by them. I'll develop the discipline according to how important they are to me.

The root of the word *discipline* is *disciple*. Now, I'm not proselytizing here, I'm just telling you the truth. A disciple is a believer, a follower. One who believes in something or someone to the degree of doing what is expected by that something or someone without being told to and without hesitation. Discipleship doesn't just happen. There has to be a reason for followers to follow a leader. The most powerful reason is that the leader cares. It may start with the leader saying he/she cares, but the power is in SHOWING you care. The more you practice caring, the more your followers will believe in you, and the more they'll follow you. Eventually, they'll become your disciples. I, myself, am a disciple of several leaders, each who has, and many still do, shown me they care. In MY language, which includes actions.

We all practice discipline on a daily basis…or not…in small ways that make us who we eventually are. As Socrates said, "We are what we repeatedly do." And what we repeatedly do is based on a discipline of some sort. Discipline, by definition, is a controlled behavior. A behavior is what we do. We, no kidding, control EVERYTHING we do. The important question in developing and practicing the RIGHT discipline is WHY? (My friend, Simon Sinek, is smiling right now!) The reason we do what we do is that it affects either us or others, positively or negatively. That's it!

Here's how I practice discipline daily, and why. I enjoy driving fast. Deb won't let me drive the way I like to, though. I really enjoyed driving on the German autobahns, especially A-65, which had no speed limit. Man, get a BMW out there and fly! Love it! I can't, however, do that in Colorado Springs. For one, it's illegal, and for the other, Deb has a gadget installed on my cars that tells her when I've gone over 80 MPH. No kidding, there's an app for that!

You see, I think of speed limits as more of a guide than a law. Except on base.

I've served my country since 1972. One of my most favorite groups of Warriors is the Cops. I still call them that, and they let me. Don't get me wrong, I love ALL Airmen, but I see the Cops every day. In fact, they, often, come by my office. Love those people! I believe in those men and women who put their lives on the line for us every day. They're the first line of defense in case we're threatened.

Anyway, as Deb and I drove on to the Air Force Academy yesterday, the first persons we saw at the gate, believe it or not, were Cops! I didn't have my ID out, so Deb asks me if I know them. As I get up to the gate, I stop, say hello to A1C P (can't pronounce his name) and A1C Fleming. Fleming comes up to me and I tell him what Deb just asked me. He laughs and says, "We ALL know you, Chief!" He made me feel special.

As we drove on the base, I made sure I kept to the speed limit. Had to! I love those two Airmen, and all of their Peeps. I WILL NOT violate the law and put them in a situation where they'd have to admonish me! I will practice the discipline of maintaining the posted speeds because I believe in them! Now, I often argue with myself, you know that internal argument we have with invisible people, that the speed limits should be higher, but until they get changed, I'm not breaking them. By the way, I also use my turn signals!

We practice discipline all the time. We just don't think about it and we, surely, don't think about why, other than we may get punished if we don't. Think about why we do what we should, especially that a violation of the right things affects others, and maybe we should respect them. Maybe respect them and believe in them.

I'll be driving to work in a couple of hours. I'll keep to the speed limits as I drive on base. I can control that behavior. It may seem small, but I know my Brother and Sister Cops will appreciate it, and me. I am a disciple of the Defenders of the Force!

We, often, talk about discipline in the sense of controlling others. Real discipline doesn't work that way. If you're using discipline to control others, that's really punishment. There's a difference between discipline and punishment. If you're striving to be a Leader of Character, know the difference, and strive for discipline.

We wonder why we have leadership problems in our culture, yet our perspective on developing followers is, mostly, based on forcing people to do what we want them to do. How's that working for you?

When we strive to develop discipline, we tend to be more effective as leaders. Effectiveness, as I said before, is, simply, getting what you want continuously. We want our followers to be great and do what they should do even when we're not around. That's the sign of a good leader.

What's going to make you a Leader of Character is having the discipline to follow the Core Values on a daily basis and modeling the right behavior according to those Values. As James Clear writes in his book titled, *Atomic Habits*, there's a difference between being in motion and taking action. "The two ideas sound similar, but they're not the same. When you're in motion, you're planning and strategizing and learning. Those are all good things, but they don't produce a result. Action, on the other hand, is the type of behavior that will deliver an outcome. If I outline twenty ideas for articles I want to write, that's motion. If I actually sit down and write an article, that's action. If I search for a better diet plan and read a few books on the topic, that's motion. If I actually eat a healthy meal, that's action.

"Motion makes you feel like you're getting things done. But really, you're just preparing to get something done. When preparation becomes a form of procrastination, you need to change something. You don't want to merely be planning. You want to be practicing. If you want to master a habit, the key is to start with repetition, not perfection. You don't need to map out every feature of a new habit. You just need to practice it."

To know but not to do is not to know.

▲

One of the ways we "discipline" our Airmen in the United States Air Force is by making sure everyone pays for the indiscretions of a few, often, one. And, man, does that ever work! NOT! I remember, not long ago, the base at which I was assigned, having a Family Day scheduled for the upcoming Friday, which would give everyone a three-day weekend. Everyone was a disciple of the Wing Commander! He was THE MAN! (He happened to be male.) We all believed in him! Until, however, comma, (Ever notice that nothing positive ever follows "however, comma"?) an Airman got caught DUI the Tuesday before. In his attempt to "discipline" the Airman, and the wing, he decided to take the Family Day away. WAIT! One person messed up and everyone has to pay?! That's PUNISHMENT! When this happens, and it still does, the intent is that everyone will understand that we're a team and that the actions of one affect everyone. We have to be wingmen and ensure everyone does what's right. You'd think we'd know better. Punishing everyone for the actions of one has never worked, and never will. Sorry, but that's the truth.

Now, I'm not advocating anarchy. What I AM advocating for is that leaders, YOU, consider developing your followers into people who follow you for the right reasons, and do the right things even when you're not around. That's discipline. And that's leading. It becomes HEIRPOWER!

Leaders of Character practice the right discipline. They model the discipline of embodying the Core Values. The more you practice those, the closer you'll get to excellence.

# TEAMWORK

Teamwork, by definition, is “a group of individuals working together to achieve their goal.” I looked it up!

I know you’ve heard it, or read it, “There is no I in TEAM!” Humbug! Re-look at the definition. A team is a group of INDIVIDUALS working together to achieve their goal. I love Phil Jackson’s quote: “The strength of the team is each individual member. The strength of each member is the team.” I think it supports my premise that the team’s strength is founded on each person doing their part. Whatever that part is. The difficult question, often, is what is that part?

When we think of a team, we sometimes think in terms of a puzzle. Most people build a puzzle, or a team, from the outside in. We set the boundaries, we frame the context, if you will, then find the puzzle pieces that fit inside. Most managers do that. It’s a common process that’s efficient. A puzzle is created with a scarcity mentality, though. There’s only so much. You can’t go beyond the barriers. Never go beyond the edges. Remember what you read about managers and leaders? Managers work on efficiency. Leaders strive for effectiveness. A puzzle mentality asks managers to consider, “Where do they fit in?”

If you seek effectiveness, if you aspire to be a leader, then you should think in terms of a mosaic. A mosaic is built from the center out. There are no boundaries. It’s created from an abundance mentality. As the leader, your question is “HOW do they fit in?” It’s a different perspective because, now you can apply all of what those individuals are empowered with to accomplish your goal, or your mission.

Working together to achieve a goal first requires a goal. When individuals have a COMMON goal, then, and only then, will they, as a team, achieve teamwork and accomplish the goal. Every one of us has been endowed by our Creator with certain unalienable rights. I read that somewhere. Teamwork requires the leader to acknowledge, understand, and empower her or his teammates to be their better angels of their nature. I read that, too. Never forget the importance each person contributes for the team to maintain its viability and its strength.

Ever notice that there's nothing we do in our lives that isn't touched by someone else? Let's start with the most foundational thing we do, breathe. Although we breathe by ourselves, someone has affected that in some way. If you live in a part of the world where smog permeates the atmosphere, you know what I mean. All of the folks who have, somehow, created the smog have affected your breathing. If, on the other hand, you live where the air is pure, those who COULD have created pollutants, but chose not to, have affected your breathing as well. What's this got to do with the virtue of teamwork? We're all in this together. But we seldom consider the effect our behavior can, and does, have on others and the world we live in.

The other day, I wore a new pair of pants. Sometime during the day, I noticed something in one of my pockets that kept bothering me. It kept scratching my leg. It was a little tag. Amazing the effect that little tag had on me. People probably wondered why I was so fidgety. I, eventually, reached into that pocket and pulled out the tag that read, "Inspected by Number 22." Okay, maybe I need to find a life, but I got to thinking, who is this famous Number 22? Is it a male or female? Old or young? Tall or short? I wondered if Number 22 has a family. How many kids? I went on and on. Number 22 affected me. After I stopped wondering about Number 22, I got to thinking that he/she had done a pretty good job of inspecting the pants I was wearing. They felt comfortable. They allowed me to do my job better. Now, Number 22 could have cheated or not taken the job of inspecting those pants seriously and affected me in a negative way. Instead, he/she made my life better that day. Number 22 did her/his part. We made it a good day together! Thank you, Number 22! Thanks for making my life better! Thanks for being on my team!

▲

I've been serving at the United States Air Force Academy since 2002. On this particular day, the Freshmen, we call them Four-degrees, are going through their Rite of Passage exercise we call Recognition. I'm out on the Terrazzo observing the Four-degrees excel at what they do best. I'm over by the Hill where they're formed up doing flutter kicks. As I'm standing there, an Upperclassman walks up to me and we start to chat. As we're talking, we both notice a Four-degree who is having a really hard time doing the kicks. So hard a time that she quits! Just stops! You can see she's given all she had, but it wasn't enough.

As the Upperclassman sees her, he says, "Excuse me, Chief," and walks toward the Four-degree. "Oh no," I'm thinking. "Bad things are about to happen to My Four-degree!"

The Upperclassman lies down next to the Four-degree and starts to do flutters kicks along with all the other Four-degrees. The Four-degree sees him, takes a deep breath, then starts to do the kicks again. They both continue until the trainer-in-charge blows his whistle to move on to the next station.

The Four-degree can barely get up. She's struggling. The Upperclassman grabs her by the arm and they both head off to the next station. Curious, I follow them to see how long the Upperclassman assists the Four-degree. He helps her for at least two more stations. I eventually can't keep up with them. They continue driving on. As Dr Ken Blanchard says, "none of us is as good as all of us."

Here are some great thoughts on excellence from my Chosen Brother, CMSgt (Ret) John G. Evalle.

One of my favorite leadership books is *Exceeding Expectations, Reflections on Leadership*, by retired Air Force General Bill Looney. I have personally found a universal truth is that all great Airmen have an internal drive to perform and produce far beyond the minimum standard. Those great Airmen are never satisfied with a "that's good enough" result. Great Airmen are internally inspired and committed to putting their very best into every task they are asked to perform. They exceed your expectations.

While I was still on active duty, I had the privilege to have had several amazing young Airmen work directly for me. Before I retired, one of these Airmen, a former Executive Assistance that worked for me for a year, asked if she could come by my office because she had something to give me. Two important things to know: first, I've shared with many Airmen my love for the leadership lessons from the Civil War and specifically the Battle of Gettysburg; and second, I had a beautiful puzzle of a Beverly Doolittle Chief painting framed and displayed in my office. My wife had this puzzle framed for me and it is very special.

A few days later, the Airman came to my office and presented me a framed puzzle of Colonel Joshua Chamberlain of the 20th Maine Volunteer Infantry Regiment during the defense of Little Round Top at the Battle of Gettysburg. I was blown away that she would take the time to find such a meaningful image for me, but even more impressive was her commitment to assemble this massive 1000-piece puzzle. It was such an amazing gift, but I could tell there was still something bothering this Airman. She then told me that she had wanted to give me this farewell gift for a while, but something happened. She confided to me that when she was first completing the puzzle, she found there to be three pieces missing. She was convinced that the missing pieces had fallen on the floor and her dogs ate them. I quickly looked back at the framed puzzle and there were no pieces missing.

I said that I didn't understand and that it looked perfect. This great Airman then shared something that I will never forget. She tells me that she purchased a second puzzle and went through the entire new puzzle box until she found the three missing pieces to complete the picture. It is hard to explain, but if you have ever assembled a large puzzle, you can appreciate the extensive task of going through 1000 puzzle pieces one after another to find the exact missing pieces. I initially thought the gift was amazing, but after hearing what happened this Airman's gift of her commitment to excellence far exceeded any physical presentation.

This great Airman had a vision and was internally inspired to make her vision happen. And when faced with a daunting challenge, she went above and beyond to complete her task.

No one would have ever known if she chose to never complete the puzzle, but she chose to do something far beyond what I would have ever expected any Airman to do.

I was very blessed to be a Chief Master Sergeant in the United States Air Force and to have worked with many great Airmen during my thirty-year career. As I think back to those great Airmen, they all shared that same trait of being internally driven to perform every task with a sense of pride and ownership.

That internal drive motivated them to exceed the expectations of their leadership. That is what great Airmen do and how they do it.

CMSgt (Ret) John G. Evalle
Former Command Chief Master Sergeant
60th Air Mobility Wing (Travis AFB, CA) and 436th Airlift Wing (Dover AFB, DE)

Life is a continuum. None of us is perfect. But that shouldn't keep us for striving for it. Chances are we'll fail. But as Zig Ziglar said, "Failure is an event, not a person." As long as we're heading in the right direction on that continuum of excellence, as long as we're ALL IN! to doing everything we do at our highest level, we'll be better.

Let me close this chapter out with an apropos thought from a great thinker, Plato, who said, "Excellence is not a gift, but a skill that takes practice. We do not act 'rightly' because we are 'excellent,' in fact, we achieve 'excellence' by acting 'rightly.'"

# FINAL THOUGHTS....

We all come to serve for our own reasons. I'm convinced that most of us stay because of the people. Airmen and their families are the best in the world. They ARE! They're GREAT! The way we ensure greatness is by practicing greatness. As Saint Teresa of Calcutta said, "We can do no great things - only small things with great love."

Small things make THE difference. Why do people follow you, because of the great things you've done? Nope. It's the small things, with great love, or, at least caring. It's about modeling. Your followers may not do what you say, but they'll do what you do.

▲

I'm on my way into a classroom with three senior cadets at the Air Force Academy. My bride has trained me to be meticulously clean so that's how I am. As my cadets and I are about to enter the classroom, I notice a tiny piece of paper on the carpet. I can't let that go. Deb would yell at me! So, I stop. My cadets stop. They look at each other with a what-just-happened look on their faces. I bend down and pick up the tiny piece of paper and put it in the nearby trash can. They watch me, then look around and notice some other pieces of paper strewn throughout the floor. They, on their own, pick up the pieces of paper and throw them into the trash can. That's how you show initiative. That's how you show others how to lead. Interestingly, from that day on, every time we entered that room, those future leaders would look around for trash before entering. This may seem small, but it's powerful. The small stuff has big impact.

Let's finish this discussion almost as I started. I started by talking about core values. The real value of this whole book is in the application of the tenets of the core values to the level that they become our CORPS values. That is, that we all apply them, do them, every day.

I've referred to Dr Covey's idea of life as a continuum. I'm convinced that that continuum starts up all over again every day. We may have done well yesterday, but that's gone. It all starts up again today. We have to keep driving. In the right direction.

You may recall my story about how I'm tested every day as I drive in to work. Well, I'm tested as well when I leave work.

The speed limit on North Gate road, which leads, incredibly, out the North Gate of the Air Force Academy, is forty-five miles per hour. I ALWAYS drive the speed limit…on base. As you approach the gate there's a bridge right before it. You're going downhill. The speed limit changes right as you pass the static B-52 on the right. It changes to thirty-five, then to twenty-five, and, finally, it's twenty miles per hour as you cross the bridge and make a couple of twists and turns, eventually leaving the base. Before I became intentional about practicing the Core Values, I did what most folks do. I paid no attention to those speed limits. In fact, I remember speeding past Chaplain Jim Glass one day, waving at him as he drove the correct speed. He gave me a look that I still recall. Forgive me, Chappy! I've repented.

I ensure that I adhere to the speed limits on base for two reasons. First, it's the right thing to do. Oh, it may seem a small thing, especially at zero-dark-thirty, when no one else is on the road, but if you're intentional about doing small things right, that practice and, maybe even, habit, will affect other, bigger, things. I stop at those stinking red lights in Monument, and I drive the speed limit on base. I'm working on driving the speed limits off base.

The other reason I follow the road rules on base is that I know all, or most, of the Security Forces Warriors, the Defenders of the Force. I often call them by an old-school term, Cops! I love those Warriors! I respect them and am grateful to them for putting their lives on the line for me and my family every day. I do my best to practice integrity on base because I don't want to put My Cops in a situation where they have to correct me, especially, legally. I would feel terrible if I put them in that situation. I won't do it to them, or to me!

One of my most favoritest Cops in the whole world is Senior Airman Maserati. That's not his real name. It's what I call him. Great young warrior! One day, as I come up to the gate there must be fifty cars backed up because the Cops are doing random inspections. They're searching every fifth car. As I come up to the gate, Maserati spots me, does some internal math, and realizes that I'm the fifth car so he'll have to pull me over and inspect my car. As he directs me off the road, he's apologizing, explaining to me why he's pulling me over. "Maserati," I tell him, "you've got a duty to fulfill. If it's to pull me over and inspect my car, that's what you do. It's MY duty to respect you for doing YOUR duty. Let's do this!" He stood a little taller as he conducted his inspection. He thanked me, smiling as big as he could, and I went on my way. I still love that young man!

▲

At the very beginning of sharing my thoughts and stories I offered that we always do what's most important to us. I'm not sure that's true anymore. Hey! It's taken me some time to write this book so I may have learned something along the way. Here's what I'm pretty sure IS true. We do what's most important to those who are most important to us. Sometimes that's ourselves. Sometimes it's others.

I stop at those lights in Monument because the people I lead and teach are important to me. I slow down as I leave the base because those who man that gate are important to me.

Just imagine that integrity, service and excellence are as important to me as they are to you. And that they're as important to you as they are to me. Imagine a world where they're important to each and all of us and that we practice them as best we can, intentionally, every day in everything we do. The mind is so powerful that we can truly, no-kidding, imagine that world. But it won't be real until we connect the heart, the core, to the hands. My last Covey quote is this, "Everything is created twice...first in the mind, and then in the physical sense." Don't just imagine it. DO IT! DO IT NOW! DO IT RIGHT! MAKE IT LAST!

I may have said this before. What I've written here shouldn't be taken as prescriptive. I'm not worthy of telling you what to do or how to do it. My purpose is to help you become better, because as good as we may be, we can always get better. My intent is to help you empower yourself to be better by inspiring you to re-commit to practicing the Air Force Core Values.

I've served in our Air Force (proudly, by the way) since 1972. One of the many things I've observed throughout my service is that we're anxious for answers. That may be okay, but if we don't ask the right QUESTIONS, we'll get great answers to the WRONG ones. I'm hoping that what you read here will inspire you to ask the critical questions of "What's MOST important to me?" and "How will I, do I, apply that on a daily basis?" I'm hoping that your answers will have some correlation to our Air Force Core Values. The key to our success as a Force is making our Core Values our Corps Values.

Day is almost done this Thanksgiving Day, 2017. I'm sitting here thinking and thanking. I'm so grateful for all that the Creator has blessed me with, knowing that I don't deserve it, but counting it as Grace.

I'm serving in my forty-seventh year of service, with a two-week break (That's how long it took me to get to USAFA from Germany.). I didn't join the Air Force because I was a patriot. I joined because I was trying to elude the draft. The Viet Nam War was still going on. I've never won anything…except the draft. I can't recall why, but I had a 4-F deferment because I was in college when, all-of-a-sudden, President Nixon decided that I was 1-A. I remember seeing the front page of our local newspaper, The El Paso Herald Post, as if I was looking at it right now. Looking at the front page, the left column noted all of the draft numbers that would be called up within the next month. Mine was about fourth or fifth. It was time to do something! And it wasn't going to be move to Canada! I had a very close friend, Chuck Wilson, who got me an audition with the Fifteenth Air Force Band at March Air Force Base. I went out, auditioned, and was selected. I ended up in the Pacific anyway, but in a blue uniform instead of green.

I think we all go through it...at about the three-year mark I was ready to get out. My initial enlistment was about to be done and I was more than ready! Oh, WAIT! I suppose things were bad because they asked me to stay. In fact, I was offered a seven-thousand-dollar bonus to stay. I was about to get promoted to Staff Sergeant as well. And in those days seven grand was a lot of money and Staff Sergeant pay wasn't too shabby either. I could handle it for four more years. NO MORE, though!

So, it's now a few decades later and I'm still in. ALL IN! There's a reason I've stayed! The Air Force has been through many changes in the past 46 years. Some of them not-so-good. But what hasn't changed is the quality of the people. Alexis de Tocqueville is credited with saying that "America is great because she is good. If America ceases to be good America will cease to be great." People argue over whether or not he said that. I don't care. The thought is sound. And here's what I say: "The United States Air Force is great because it is good. If it ceases to be good, it will cease to be great!" What makes the Air Force great is the great people who serve as Airmen in the Profession of Arms!

I'm blessed with the honor of being among the best of the best on a daily basis, whether it's at the Air Force Academy or on any other Air Force Base in the world. People constantly ask me, "How ya doin, Chief?!" I consistently answer, "GREAT! I hang with the best people in the world. I can't help but be GREAT!" I say that with all honesty and feel it whole-heartedly.

The question is, how do we maintain our greatness as a force? The answer is we embody our Core Values. We embody them and apply them until they become our Corps Values, our identity. Until we can't help but uphold them. We do what's right. We serve others. We're ALL IN! ALL THE TIME!, in everything we do.

I salute you for taking the time to read my thoughts and for practicing what you know you should. I look forward to serving with you for a few more years. Until I punch, I will serve you in any way that I can. I call it HEIRPOWER!

There was a farmer who grew excellent quality wheat and every season he won the award for the best grown in his county. One year a reporter from the local newspaper interviewed the farmer and learned that each Spring the man shared his seed with his neighbors so that they too could plant it in their fields...

“How can you afford to share your best wheat seed with your neighbors when they are entering their crops in the competition with yours?" the reporter asked.

“Why that's very simple,” the farmer explained. "The wind picks up pollen from the developing wheat and carries it from field to field. If my neighbors grow inferior wheat, cross-pollination will steadily degrade the quality of all the wheat, including mine. If I am to grow good wheat, I must help my neighbors grow good wheat."

The reporter realized how the farmer's explanation also applied to peoples' lives in the most fundamental way. Those who want to live meaningfully and well must help enrich the lives of others, for the value of a life is measured by the lives it touches. And those who choose to be happy must help others find happiness, for the welfare of each is bound up with the welfare of all.

# ABOUT THE AUTHOR

Bob Vásquez is currently a curriculum developer at the United States Air Force Academy. He previously served as the course director for the freshman seminar offered through the US Air Force Academy's Center for Character Development. He served in the Air Force for more than 30 years before retiring on 31 October 2002. Bob invested 24 of those years in Air Force bands throughout the world. He also served as Commandant of the Noncommissioned Officer Academy at March Air Force Base, California; Senior Enlisted Advisor to the Commander of the 92nd Airlift Wing at Fairchild Air Force Base, Washington; Deputy Director of the Family Support Center at Ramstein Air Base, Germany; and Superintendent for the 86th Support Group at Ramstein.

A proud member of the Air Force Academy team since 16 November 2002, Bob is also an adjunct professor at the University of Colorado at Colorado Springs.

He's the author of *Heirpower! Eight Basic Habits of Exceptionally Powerful Lieutenants*, *So Ya Wanna Be THE Chief?!*, *S.S.G.T!*, *The College Freshman's Beyond Survival Guide*, *What I Learned from Dad Made Me a Better Man!*, *CAPstone!, and Another Shade of Blue: Stories by Hispanic United States Air Force Airmen.*

Although a musician, motivational speaker, life coach, and mentor, Bob considers his greatest accomplishments the successful raising of his daughters, Tesa and Elyse, six grandchildren, two sons-in-law, and growing closer to Debbie, his lovely bride of more than 43 wonderful and fulfilling years.

Connect with Bob at
bobvasquez@bobvasquez.com
(More books by Bob Vasquez on the following page.)

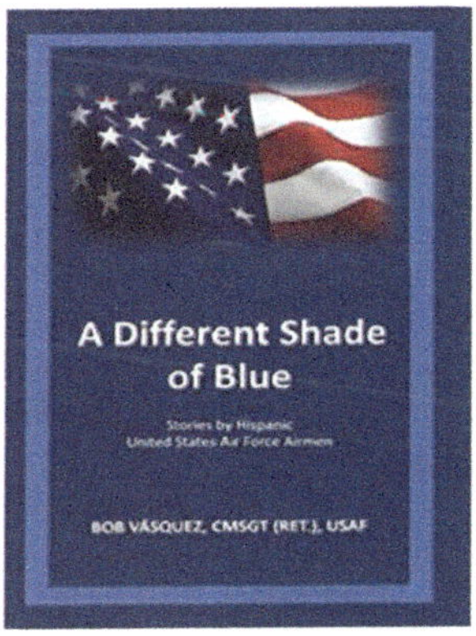
A Different Shade of Blue
Stories by Hispanic United States Air Force Airmen
BOB VÁSQUEZ, CMSGT (RET.), USAF

The POWER of SUPERvision!
bob vásquez

S.S.G.T.
Four Leadership Lessons for New NCOs!
BOB VÁSQUEZ, CMSGT (RET), USAF

So Ya Wanna Be THE Chief?!
Twelve Lessons in Servant-Leadership
Bob Vasquez, CMSgt (Ret), USAF

CAPstone!
10 Habits of Highly Effective Cadet Leaders
A Leadership Book For Cadets, By Cadets

The College Freshman BEYOND Survival Guide!
Bob Vásquez
Elyse Vásquez White